LONDON MATHEMATICAL SOCIETY LECTURE NOTE SERIES

The books in the series listed below are available from booksellers, or, in case of difficulty, from Cambridge University Press.

34 Representation theory of Lie groups, M.F. ATIYAH *et al*
36 Homological group theory, C.T.C. WALL (ed)
39 Affine sets and affine groups, D.G. NORTHCOTT
40 Introduction to H_p spaces, P.J. KOOSIS
46 p-adic analysis: a short course on recent work, N. KOBLITZ
49 Finite geometries and designs, P. CAMERON, J.W.P. HIRSCHFELD & D.R. HUGHES (eds)
50 Commutator calculus and groups of homotopy classes, H.J. BAUES
57 Techniques of geometric topology, R.A. FENN
59 Applicable differential geometry, M. CRAMPIN & F.A.E. PIRANI
62 Economics for mathematicians, J.W.S. CASSELS
66 Several complex variables and complex manifolds II, M.J. FIELD
69 Representation theory, I.M. GELFAND *et al*
74 Symmetric designs: an algebraic approach, E.S. LANDER
76 Spectral theory of linear differential operators and comparison algebras, H.O. CORDES
77 Isolated singular points on complete intersections, E.J.N. LOOIJENGA
78 A primer on Riemann surfaces, A.F. BEARDON
79 Probability, statistics and analysis, J.F.C. KINGMAN & G.E.H. REUTER (eds)
80 Introduction to the representation theory of compact and locally compact groups, A. ROBERT
81 Skew fields, P.K. DRAXL
82 Surveys in combinatorics, E.K. LLOYD (ed)
83 Homogeneous structures on Riemannian manifolds, F. TRICERRI & L. VANHECKE
86 Topological topics, I.M. JAMES (ed)
87 Surveys in set theory, A.R.D. MATHIAS (ed)
88 FPF ring theory, C. FAITH & S. PAGE
89 An F-space sampler, N.J. KALTON, N.T. PECK & J.W. ROBERTS
90 Polytopes and symmetry, S.A. ROBERTSON
91 Classgroups of group rings, M.J. TAYLOR
92 Representation of rings over skew fields, A.H. SCHOFIELD
93 Aspects of topology, I.M. JAMES & E.H. KRONHEIMER (eds)
94 Representations of general linear groups, G.D. JAMES
95 Low-dimensional topology 1982, R.A. FENN (ed)
96 Diophantine equations over function fields, R.C. MASON
97 Varieties of constructive mathematics, D.S. BRIDGES & F. RICHMAN
98 Localization in Noetherian rings, A.V. JATEGAONKAR
99 Methods of differential geometry in algebraic topology, M. KAROUBI & C. LERUSTE
100 Stopping time techniques for analysts and probabilists, L. EGGHE
101 Groups and geometry, ROGER C. LYNDON
103 Surveys in combinatorics 1985, I. ANDERSON (ed)
104 Elliptic structures on 3-manifolds, C.B. THOMAS
105 A local spectral theory for closed operators, I. ERDELYI & WANG SHENGWANG
106 Syzygies, E.G. EVANS & P. GRIFFITH

107 Compactification of Siegel moduli schemes, C-L. CHAI
108 Some topics in graph theory, H.P. YAP
109 Diophantine analysis, J. LOXTON & A. VAN DER POORTEN (eds)
110 An introduction to surreal numbers, H. GONSHOR
111 Analytical and geometric aspects of hyperbolic space, D.B.A.EPSTEIN (ed)
113 Lectures on the asymptotic theory of ideals, D. REES
114 Lectures on Bochner-Riesz means, K.M. DAVIS & Y-C. CHANG
115 An introduction to independence for analysts, H.G. DALES & W.H. WOODIN
116 Representations of algebras, P.J. WEBB (ed)
117 Homotopy theory, E. REES & J.D.S. JONES (eds)
118 Skew linear groups, M. SHIRVANI & B. WEHRFRITZ
119 Triangulated categories in the representation theory of finite-dimensional algebras, D. HAPPEL
121 Proceedings of *Groups - St Andrews 1985*, E. ROBERTSON & C. CAMPBELL (eds)
122 Non-classical continuum mechanics, R.J. KNOPS & A.A. LACEY (eds)
124 Lie groupoids and Lie algebroids in differential geometry, K. MACKENZIE
125 Commutator theory for congruence modular varieties, R. FREESE & R. MCKENZIE
126 Van der Corput's method of exponential sums, S.W. GRAHAM & G. KOLESNIK
127 New directions in dynamical systems, T.J. BEDFORD & J.W. SWIFT (eds)
128 Descriptive set theory and the structure of sets of uniqueness, A.S. KECHRIS & A. LOUVEAU
129 The subgroup structure of the finite classical groups, P.B. KLEIDMAN & M.W.LIEBECK
130 Model theory and modules, M. PREST
131 Algebraic, extremal & metric combinatorics, M-M. DEZA, P. FRANKL & I.G. ROSENBERG (eds)
132 Whitehead groups of finite groups, ROBERT OLIVER
133 Linear algebraic monoids, MOHAN S. PUTCHA
134 Number theory and dynamical systems, M. DODSON & J. VICKERS (eds)
135 Operator algebras and applications, 1, D. EVANS & M. TAKESAKI (eds)
136 Operator algebras and applications, 2, D. EVANS & M. TAKESAKI (eds)
137 Analysis at Urbana, I, E. BERKSON, T. PECK, & J. UHL (eds)
138 Analysis at Urbana, II, E. BERKSON, T. PECK, & J. UHL (eds)
139 Advances in homotopy theory, S. SALAMON, B. STEER & W. SUTHERLAND (eds)
140 Geometric aspects of Banach spaces, E.M. PEINADOR and A. RODES (eds)
141 Surveys in combinatorics 1989, J. SIEMONS (ed)
142 The geometry of jet bundles, D.J. SAUNDERS
143 The ergodic theory of discrete groups, PETER J. NICHOLLS
144 Introduction to uniform spaces, I.M. JAMES
145 Homological questions in local algebra, JAN R. STROOKER
146 Cohen-Macaulay modules over Cohen-Macaulay rings, Y. YOSHINO
147 Continuous and discrete modules, S.H. MOHAMED & B.J. MÜLLER
148 Helices and vector bundles, A.N. RUDAKOV et al
149 Solitons, nonlinear evolution equations and inverse scattering, M.A. ABLOWITZ & P.A. CLARKSON
150 Geometry of low-dimensional manifolds 1, S. DONALDSON & C.B. THOMAS (eds)
151 Geometry of low-dimensional manifolds 2, S. DONALDSON & C.B. THOMAS (eds)
152 Oligomorphic permutation groups, P. CAMERON
153 L-functions and arithmetic, J. COATES & M.J. TAYLOR (eds)
154 Number theory and cryptography, J. LOXTON (ed)
155 Classification theories of polarized varieties, TAKAO FUJITA
156 Twistors in mathematics and physics, T.N. BAILEY & R.J. BASTON (eds)
157 Analytic pro-*p* groups, J.D. DIXON, M.P.F. DU SAUTOY, A. MANN & D. SEGAL
158 Geometry of Banach spaces, P.F.X. MÜLLER & W. SCHACHERMAYER (eds)
159 Groups St Andrews 1989 Volume 1, C.M. CAMPBELL & E.F. ROBERTSON (eds)
160 Groups St Andrews 1989 Volume 2, C.M. CAMPBELL & E.F. ROBERTSON (eds)
161 Lectures on block theory, BURKHARD KÜLSHAMMER
162 Harmonic analysis and representation theory for groups acting on homogeneous trees, A. FIGA-TALAMANCA & C. NEBBIA

London Mathematical Society Lecture Note Series, 162

Harmonic Analysis and Representation Theory for Groups Acting on Homogeneous Trees

Alessandro Figà-Talamanca
Department of Mathematics, University of Rome "La Sapienza"
and
Claudio Nebbia
Department of Mathematics, University of Rome "La Sapienza"

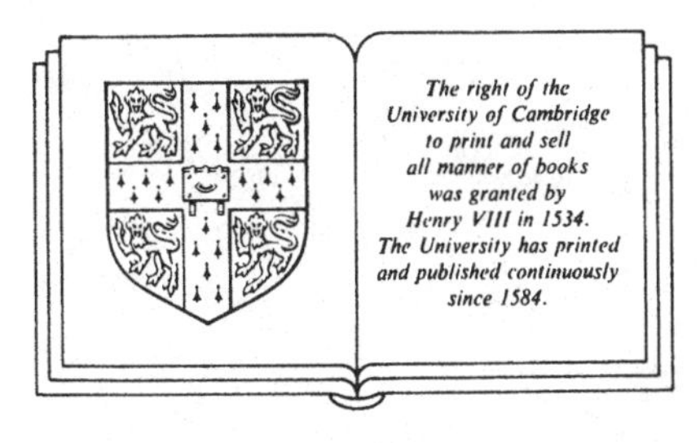

CAMBRIDGE UNIVERSITY PRESS
Cambridge
New York Port Chester Melbourne Sydney

CAMBRIDGE UNIVERSITY PRESS
Cambridge, New York, Melbourne, Madrid, Cape Town, Singapore, São Paulo

Cambridge University Press
The Edinburgh Building, Cambridge CB2 8RU, UK

Published in the United States of America by Cambridge University Press, New York

www.cambridge.org
Information on this title: www.cambridge.org/9780521424448

First published 1991
Re-issued in this digitally printed version 2007

A catalogue record for this publication is available from the British Library

ISBN 978-0-521-42444-8 paperback

CONTENTS

PREFACE

Over the past few years, we ran a Seminar in Harmonic Analysis at the Mathematics Department of the University of Rome "La Sapienza". In this seminar many of the talks given by staff members and visitors were concerned, directly or indirectly, with infinite trees or tree-like graphs, and their automorphism groups. Seminar notes were occasionally taken by one or both of us, and sometimes written up informally for distribution to newcomers to the seminar. After a while, we felt that it would be convenient to give a more coherent organization to these notes. Once this decision was taken it became apparent that, at the cost of some omission, the general aim of describing the group of automorphisms of a homogeneous tree and its irreducible unitary representations would provide a convenient focus which would include much of the material we had in mind. We felt that this approach would shed light on the connection between harmonic analysis on trees and harmonic analysis on hyperbolic spaces, by emphasizing the strict analogy between the group of automorphisms of the tree and real rank 1 semisimple Lie groups. This choice left out a lot of valuable material specifically concerning free groups and free products of finite groups. We felt however that the notes **[F-T P2]** and the memoir **[F-T S2]** could provide an introduction to these topics. We also decided not to treat the case of a semihomogeneous tree. Semihomogeneous trees are natural objects because they are exactly the Bruhat-Tits buildings of rank 1 [BT]. It follows that every rank 1 reductive algebraic group over a local field is a closed subgroup of the group of automorphisms of a semihomogeneous tree. We felt however that, while no major conceptual step is needed to extend the theory of representations of the group of automorphisms of a homogeneous tree to the case of a semihomogeneous tree, from a practical point of view the notation would have become more burdensome. The connection with matrix groups over local fields is explained in these notes by giving, in the Appendix, Serre's construction of the tree of **PGL**$(2,\mathfrak{F})$ where $\mathfrak{F}$ is a local field.

Chapter I contains a description of the geometry of a homogeneous tree $\mathfrak{X}$ and its boundary, the group of automorphisms Aut$(\mathfrak{X})$ and some of its notable subgroups. Chapter II contains the boundary theory for eigenfunctions of the Laplace (or Hecke) operator on the tree and a complete description of spherical functions and spherical representations which applies to every closed subgroup of Aut$(\mathfrak{X})$ with transitive action on the vertices and the boundary of the tree. It also contains the proof of an important result due to T. Steger which asserts that every spherical representation (with one possible exception) of Aut$(\mathfrak{X})$ restricts irreducibly to any cocompact discrete subgroup. Chapter III contains a description of square-integrable representations of Aut$(\mathfrak{X})$ following the beautiful geometric classification due to G.I. Ol'shianskii. At the end of the chapter we give the complete Plancherel formula for Aut$(\mathfrak{X})$.

The Appendix, as already mentioned, contains a complete and elementary account of the construction of the tree of **PGL**$(2,\mathfrak{F})$ and a discussion of the action of this group on its tree.

These notes owe a great deal to the many friends, colleagues and students who participated in our seminar. We would like to thank especially M.G. Cowling, F.I. Mautner and R. Szwarc for the many critical observations and comments which stimulated our work and often found their way into these notes.

Very special thanks are due to Tim Steger who contributed in many ways with help and advice in the preparation of these notes. He also gave his permission to include in these notes his yet unpublished restriction theorem, which was presented at our seminar in 1987.

CHAPTER I

1. Graphs and trees. A *tree* is a connected graph without circuits. This definition requires a word of explanation of the terms *graph*, *connected*, and *circuit*. A graph is a pair $(\mathfrak{X},\mathfrak{E})$ consisting of a set of *vertices* $\mathfrak{X}$ and a family $\mathfrak{E}$ of two-element subsets of $\mathfrak{X}$, called *edges*. When two vertices x, y belong to the same edge (i.e., $\{x,y\}\in\mathfrak{E}$) they are said to be *adjacent*; we also say that x and y are *nearest neighbors*.

A *path* in the graph $(\mathfrak{X},\mathfrak{E})$ is a finite sequence $x_0,\ldots,x_n$, such that $\{x_i,x_{i+1}\}\in\mathfrak{E}$. A graph is called *connected* if, given two vertices $x,y\in\mathfrak{X}$, there exists a path $x_0,\ldots,x_n$, with $x_0=x$ and $x_n=y$.

A *chain* is a path $x_0,\ldots,x_n$, such that $x_i\neq x_{i+2}$, for $i=0,\ldots,n-2$. A chain $x_0,\ldots,x_n$, with $x_n=x_0$ is called a *circuit*. In particular if $(\mathfrak{X},\mathfrak{E})$ is a tree and $x,y\in\mathfrak{X}$, there exists a unique chain $x_0,\ldots,x_n$, joining x to y. We denote this chain by $[x,y]$.

We are interested in *locally finite* trees. These are trees such that every vertex belongs to a finite number of edges. The number of edges to which a vertex x of a locally finite tree belongs is called the *degree* of x. If the degree is independent of the choice of x, then the tree is called *homogeneous*. In these notes we will treat mainly *locally finite homogeneous trees*.

The common degree of all vertices of a homogeneous tree is called the *degree of the tree* and is generally denoted by q+1. The reason for this notation is that, as will be shown in the Appendix, the number q may be identified, in many cases, with the order of a certain finite field. Furthermore many of the formulae appearing in the sequel, and especially in the explicit computation of spherical functions (Chapter II, below) involve powers of q, rather than $q+1$.

There are also nonhomogeneous trees which are of interest. An important example is that of a *semihomogeneous tree*. Suppose that l and q are positive integers. A tree such that every vertex has degree $l+1$ or $q+1$, and such that two adjacent vertices have different degrees, is called *semihomogeneous* of degree (l,q).

The set of vertices of a homogeneous or semihomogeneous tree is always infinite. A tree may be represented graphically as shown in Figs 1 and 2.

$q+1 = 3$ $q+1 = 4$

Fig.1 Homogeneous trees

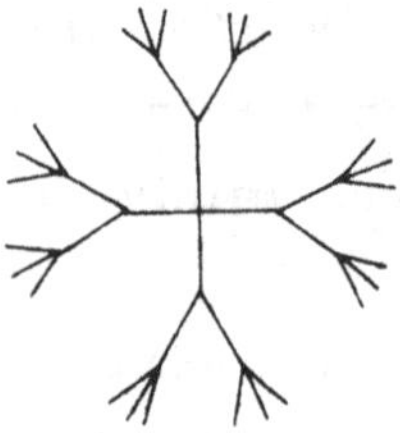

Fig.2. Semihomogeneous tree: $q = 2$, $l = 3$

The set of vertices of a tree is naturally a metric space.

The *distance* $d(x,y)$ between any two distinct vertices x and y is defined as the number of edges in the chain $[x,y]$ joining x and y, in other words the *length* of $[x,y]$.

The metric space structure of the set of vertices $\mathfrak{X}$ suffices to define the tree uniquely because two vertices belong to the same edge if and only if their distance is 1. We will often think of a tree as a set of vertices with a metric which makes it into a tree.

An *infinite chain* is an infinite sequence $x_0, x_1, x_2, \ldots,$ of vertices such that, for every i, $x_i \neq x_{i+2}$ and $\{x_i, x_{i+1}\}$ is an edge.

We define an equivalence relation on the set of infinite chains, by declaring two chains $x_0, x_1, x_2, \ldots$ and $y_0, y_1, y_2, \ldots$ equivalent if (as sets of vertices) they have an infinite intersection. This means that there is an integer $n \in \mathbb{Z}$ such that $x_k = y_{k+n}$ for every k sufficiently large. The *boundary* Ω of a tree is the set of equivalence classes of infinite chains. Observe that an infinite chain identifies uniquely a point of the boundary, which may be thought of as a point at infinity. Sometimes the points of the boundary are called *ends* of the tree.

An alternative way to define the boundary is by fixing a vertex x_0 and considering all infinite chains which start at x_0. A boundary point is associated with a unique infinite chain starting at x_0.

A *doubly infinite chain* is a sequence of vertices indexed by the integers, $\ldots x_{-2}, x_{-1}, x_0, x_1, x_2, \ldots,$ with the properties that $x_i \neq x_{i+2}$, and $\{x_i, x_{i+1}\}$ is an edge for every integer $i \in \mathbb{Z}$. A doubly infinite chain is also called an *infinite geodesic*. It identifies two boundary points. Conversely, given two distinct boundary points $\omega_1, \omega_2 \in \Omega$, there is a unique geodesic joining them. We denote this geodesic by (ω_1, ω_2). We also use the notation $[x, \omega)$ for an infinite chain starting at x in the direction of ω (that is belonging to the equivalence class ω).

Since we also want to consider the direction of chains, infinite chains and geodesics, $(\omega, x]$ will be considered different from $[x, \omega)$ even though they have the same vertices, and similarly $[x, y]$ is formally different from $[y, x]$ and (ω_1, ω_2) is different from (ω_2, ω_1).

All these concepts are more or less geometrically evident and may be illustrated with a picture

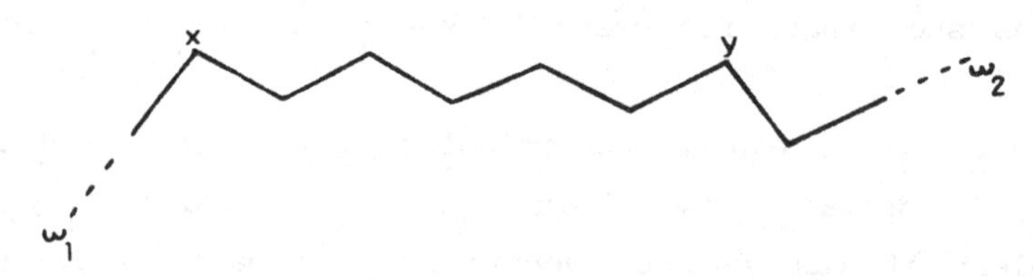

Fig.3 The boundary points ω_1 and ω_2 identify a geodesic (ω_1, ω_2)

The space $\mathfrak{X}\cup\Omega$ can be given a topology in which $\mathfrak{X}\cup\Omega$ is compact, the points of $\mathfrak{X}$ are open and $\mathfrak{X}$ is dense in $\mathfrak{X}\cup\Omega$. To define this topology it suffices to define a basis of neighborhoods for each boundary point (because each vertex is open). Let $\omega\in\Omega$, and let x be a vertex. Let $\gamma=[x, \omega)$ be the infinite chain from x to ω. For each $y\in[x, \omega)$ the neighborhood $\mathfrak{C}(x, y)$ of ω is defined to consist of all vertices and all end points of the infinite chains which include y but no other vertex of $[x, y]$ (Fig.4).

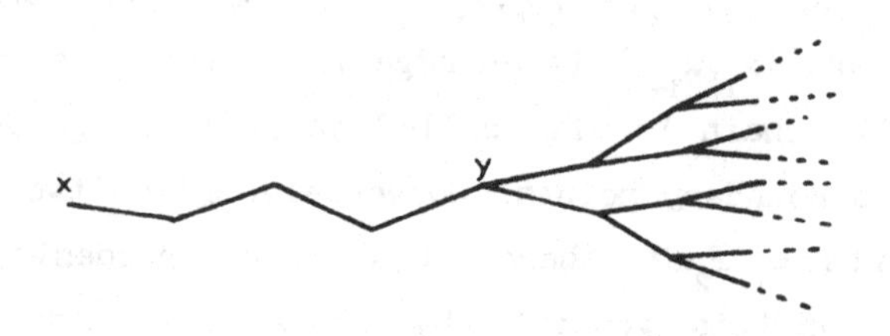

Fig.4

It is not difficult to show that $\mathfrak{X}\cup\Omega$ is compact and $\mathfrak{X}$ is dense in Ω. The relative topology on Ω (under which Ω is compact) is best described by the open sets $\Omega(x,y)$ consisting of all boundary points associated to infinite chains which start at x and pass through y (in this order). In this way, for each vertex x, $\Omega(x,x)=\Omega$ and, for every positive integer n, $\Omega=\cup\{\Omega(x,y)\colon \mathbf{d}(x,y)=n\}$. Thus the family $\{\Omega(x,y)\colon \mathbf{d}(x,y)=n\}$ is a partition of Ω into $(q+1)q^{n-1}$ open and compact sets, where q+1 is the degree of the tree. Using these partitions we can define a measure ν_x on the algebra of sets generated by the sets $\Omega(x,y)$, by letting $\nu_x(\Omega(x,y))=1/(q+1)q^{n-1}$, if $\mathbf{d}(x,y)=n$. The positive measure ν_x may be extended to a Borel probability measure on Ω.

2. The free group as a tree. We preview in this section an example to which we will return in Section 7. Let $\mathbb{F}_2$ be the free group with two generators a and b. An element of $\mathbb{F}_2$ is a reduced word in the letters a, a^{-1}, b, b^{-1}. We denote by e the empty word, which is the identity of $\mathbb{F}_2$. There is a natural correspondence between $\mathbb{F}_2$ and the vertices of a tree of degree 4, which is obtained by defining two words x and y to be in the same edge if $y^{-1}x$ is one of the generators or their inverses. This means that x and y can be obtained one from the other by right multiplication with an element of $\{a, a^{-1}, b, b^{-1}\}$. The tree which is so defined is described by Fig.5

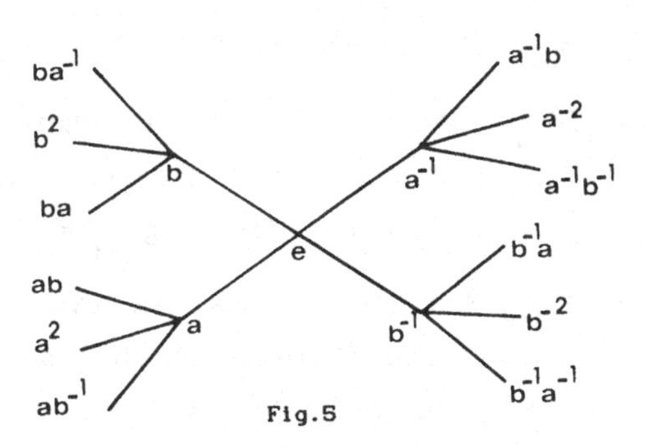

Fig.5

Other groups which may be similarly associated with homogeneous trees are described in Section 6, as certain discrete subgroups of the group of automorphisms of a tree.

It is interesting that the boundary of the tree of Fig.5 can be identified with the set of *infinite reduced words* in the letters $\{a,b,a^{-1},b^{-1}\}$. That is the words $\omega=x_1x_2x_3\ldots$, with $x_i\neq x_{i-1}^{-1}$ and $x_i\in\{a,b,a^{-1},b^{-1}\}$. We may also observe that left multiplication by elements of $\mathbb{F}_2$ on itself gives rise to an isometry of the tree. Left multiplication by a finite word is also defined on the infinite words and gives rise to a homeomorphism of Ω (see Section 7, below).

3. Automorphisms of a tree. An automorphism of a tree is a bijective map of the set of vertices onto itself which preserves the edges. An automorphism is also an isometry of the metric space $\mathfrak{X}$ endowed with the natural metric. Conversely, every isometry of $\mathfrak{X}$ is also an automorphism. We shall give presently a description of the automorphisms of a tree. We first need a preliminary result.

(3.1) *LEMMA. Let* g *be an automorphism of the tree and* x *a vertex. Let* $x=x_0,x_1,\ldots,x_n=g(x)$ *be the chain joining* x *to* $g(x)$, *and suppose that* $n>0$. *If* $g(x_1)\neq x_{n-1}$, *then there exists a doubly infinite chain,*

$$\{\ldots x_{-n},x_{-n+1},\ldots,x_{-1},x_0,x_1,\ldots,x_n=g(x_0),\ x_{n+1},\ldots\},$$

such that $g(x_j)=x_{j+n}$, *for every* $j\in\mathbb{Z}$.

PROOF. If $g(x_1)\neq x_{n-1}$ we can extend the chain $[x_0,g(x_0)]$ by defining $x_{n+1}=g(x_1)$. Since g is bijective, $g(x_0)=x_n$ implies that $g(x_2)\neq x_n$. Therefore we can define $g(x_2)=x_{n+2}$. By induction we define $g(x_k)=x_{j+k}$ and similarly $x_{j-k}=g^{-1}(x_{j-k})$. We obtain in this way a doubly infinite chain on which g acts according to the formula $g(x_j)=x_{n+j}$ for every $j\in\mathbb{Z}$. ∎

(3.2) *THEOREM. Let* g *be an automorphism of a tree; then one and only one of the following occurs.*

(1) g *stabilizes a vertex.*

(2) g *stabilizes an edge exchanging the vertices of the same edge.*

(3) *There exist a doubly infinite chain* $\gamma=\{x_n\}$ *and an integer* j *such that* $g(x_n)=x_{n+j}$ *for every* $n\in\mathbb{Z}$.

PROOF. Let g be any automorphism and let $j=\min\{\mathbf{d}(x,g(x)):\ x\in\mathfrak{X}\}$. Let $x\in\mathfrak{X}$ be such that $\mathbf{d}(x,g(x))=j$. If $j=0$, then $g(x)=x$ and g satisfies condition (1). If $j=1$, then $\{x,g(x)\}$ is an edge. In this case g satisfies (2) if $g^2(x)=x$, or (3) if $g^2(x)\neq x$ (this follows from (3.1)). Finally if $j>1$ and $[x,g(x)]=\{x_0=x, x_1,\dots,x_j=g(x)\}$, then $g(x_1)\neq x_{j-1}$, because $\mathbf{d}(x_1,x_{j-1})=(j-2)<j$. Therefore g satisfies condition (3). The fact that only one of the three conditions holds follows readily. ∎

On the basis of (3.2) it is natural to give the following definition.

(3.3) *DEFINITION. An automorphism of a tree is called a rotation if it stabilizes a vertex; an inversion if it satisfies condition* (2) *of* (3.2); *and a translation of step* j *along* γ *if it satisfies condition* (3) *of* (3.2). *In this last case* $j=\min\{\mathbf{d}(x,g(x)):\ x\in\mathfrak{X}\}$ *and* $\gamma=\{x\in\mathfrak{X}\ :\mathbf{d}(x,g(x))=j\}$.

If the tree is homogeneous there are always rotations, inversions and translations of any step on any geodesic. For a locally finite nonhomogeneous tree the situation may be quite different. First of all the existence of a translation implies the existence of an infinite geodesic. The only possible automorphisms of a finite tree are rotations and inversions. Furthermore, for any automorphism g, the degree of x and g(x) is the same. This implies, for instance, that in a semihomogeneous tree $g(a)\neq b$ if $\{a,b\}$ is an edge. In other words every automorphism of a semihomogeneous tree is either a rotation or a translation of even step.

We remark that, if g and g' are automorphisms of a tree, then g' and $gg'g^{-1}$ are both rotations, both inversions or both translations. It is also obvious that an m-power of a step-j translation on γ is a step-$|mj|$ translation on γ, for every $m\in\mathbb{Z}$.

We conclude with a technical result which will be useful later.

(3.4) *PROPOSITION.* (1) *The composition of two inversions on distinct edges is a translation of even step.*
(2) *The composition of an inversion about an edge and a rotation which does not fix both vertices of the edge is a translation of odd step on a geodesic containing that edge.*
(3) *Every automorphism of a homogeneous tree of degree* $q+1>2$ *is a product of translations.*
PROOF. Let {a, b} and {c, d} be distinct edges, that is, having at most one vertex in common (Fig.6).

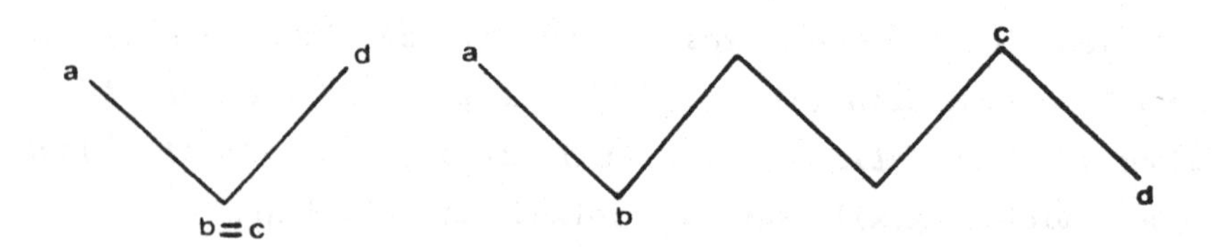

Fig.6

Let h be an inversion on {a, b} and g be an inversion on {c, d}. If b=c, then $gh(a)=g(b)=d$ and $gh(b)=g(a)\neq b$ (because $a\neq d= g^{-1}(b)$). This means that gh(b) is a vertex of distance 1 from d which is not b. By (3.1) gh is a step-2 translation on a geodesic γ containing {a, b, d}. Suppose that the two edges {a, b} and {c, d} have no vertex in common. By naming the four vertices appropriately we may suppose that $\mathbf{d}(b,c)=n$ is the distance of the two edges. This means that the chain $[a,d]$ contains $[b,c]$. Then $g(b)\notin[b,d]$ and $g(a)\notin[b,g(b)]$ (Fig.6). This implies by (3.1) that gh is a translation on a geodesic containing $[a,g(a)]$. Finally we observe that $\mathbf{d}(d,g(b))=\mathbf{d}(g(c),g(b))=n$ and therefore $\mathbf{d}(a,gh(a))=\mathbf{d}(a,g(b))=\mathbf{d}(a,b)+\mathbf{d}(b,c)+\mathbf{d}(c,d)+\mathbf{d}(d,g(b))=2(n+1)$, which means that gh is of even step. To prove (2) let g be an inversion on the edge $\{a,b\}$ and k a rotation. If $k(a)=a$ and $k(b)\neq b$, then $kg(b)=k(a)=a$, while $kg(a)=k(b)\neq b$. By (3.1), applied to $[a,b]$, we conclude that kg is a translation of step 1. If $k(a)\neq a$, and $k(b)\neq b$, let x be the point of minimal distance from $\{a,b\}$ such that $k(x)=x$. Suppose that $\mathbf{d}(x,b)=\mathbf{d}(x,a)+1$. Then $\mathbf{d}(k(a),b)=\mathbf{d}(k(a),a)+1=2\mathbf{d}(x,a)+1$. Now $k(g(b))=k(a)$ and $kg(a)=k(b)\notin[b,k(a)]$. Therefore by (3.1) applied to $[b,k(a)]$ we conclude that kg is a translation of step $2\mathbf{d}(x,a)+1$ (Fig.7).

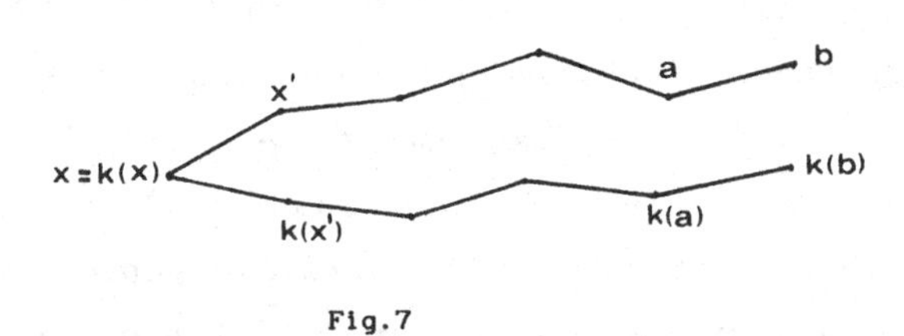

Fig.7

This proves (2). Finally assume that the tree is homogeneous with degree $q+1>2$. Let g be an inversion on the edge $\{a,b\}$ and let γ be a doubly infinite geodesic containing a but not b. (such a γ exists because $q+1>2$). Let τ be a step-1 translation on γ. Then $\tau(b)\neq a$ and $\tau g(a)=\tau(b)$ (Fig.8).

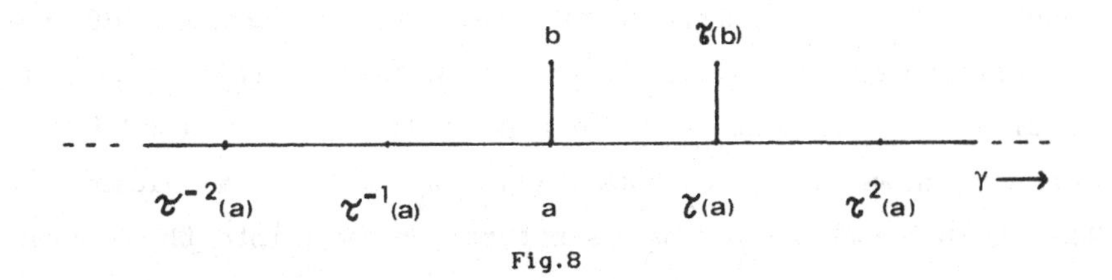

Fig.8

We apply now (3.1) to $[b,\tau(a)]$ to conclude that τg is a step-1 translation. This shows that $g=\tau^{-1}\tau g$ is the composition of two translations. Let now k be a nontrivial rotation. Since k is nontrivial and fixes a point x_0, it maps a finite chain $\mathfrak{C}$ starting at x_0 into a different finite chain $k(\mathfrak{C})$, also starting at x_0. Let a be the first point of $\mathfrak{C}$ such that $a\neq k(a)=b$ and let x be the last point of $\mathfrak{C}$ such that $k(x)=x$; then $\mathbf{d}(x,a)=\mathbf{d}(x,b)=1$. Let g be an inversion on $\{x,a\}$. Then $gk^{-1}(b)=x$ and $gk^{-1}(x)=a$, that is $\tau=gk^{-1}$ is a step-1 translation. Therefore $k=\tau^{-1}g$ is the product of a translation and an inversion. Since every inversion is the product of translations, (3) follows. ∎

4. The group of automorphisms Aut($\mathfrak{X}$). We assume from now on that $(\mathfrak{X},\mathfrak{E})$ is a homogeneous tree of degree $q+1$, and we let Aut($\mathfrak{X}$) denote the group of automorphisms of $(\mathfrak{X},\mathfrak{E})$, which is the same as the group of isometries of $\mathfrak{X}$ as a metric space.

It is not difficult to define on Aut($\mathfrak{X}$) a *locally compact topology* under which the group operations are continuous. To define a basis of neighborhoods of $g\in\mathrm{Aut}(\mathfrak{X})$, let F be a finite subset of $\mathfrak{X}$, and let $U_F(g)=\{h\in\mathrm{Aut}(\mathfrak{X})\colon g(x)=h(x), \text{ for all } x\in F\}$. It is clear that, under the topology generated by the sets $U_F(g)$, the group operations of Aut($\mathfrak{X}$) are continuous. It is also not difficult to show that the topology is *locally compact*. Indeed, for $x\in\mathfrak{X}$, let $K_x=\{g\in\mathrm{Aut}(\mathfrak{X})\colon g(x)=x\}$. By definition K_x is open. But K_x is also compact as will be presently shown. Every $g\in K_x$ acts as a permutation on the set $\mathfrak{B}_n=\{w\in\mathfrak{X}\colon \mathbf{d}(x,w)=n\}$, the set of vertices of distance n from x. This set $\mathfrak{B}_n$ has $r_n=(q+1)q^{n-1}$ elements. Therefore K_x may be thought of as a subgroup of the infinite product of the permutation groups $S(r_n)$. We will show that in this product K_x is closed. An element $g\in\prod S(r_n)$ is in the complement of K_x if, for some n, and some $w_1,w_2\in\mathfrak{B}_n$, $g(w_1)=w_2$, while some element in the chain between x and w_1 is not mapped by g into the element of the chain between x and w_2 which has the same distance from x. This condition defines an open subset of the product $\prod S(r_n)$ and therefore the complement of K_x is open. This shows that K_x is compact. As x describes $\mathfrak{X}$, the group K_x describes a family of compact open subgroups of Aut($\mathfrak{X}$) all conjugate to each other: if $g(x)=y$, then $K_y=gK_xg^{-1}$. We also observe that the group K_x is totally disconnected.

We will now prove that, if $\mathbf{d}(x,y)=n$, then $K_x \cap K_y$ has index $r_n=(q+1)q^{n-1}$ in K_x and in K_y. Indeed, let $\mathfrak{B}_n$ be as above, and for each $w \in \mathfrak{B}_n$ choose $g_w \in K_x$ such that $g_w(w)=y$. Then $K_x=\bigcup\{g_w(K_x \cap K_y)\colon w \in \mathfrak{B}_n\}$, and the cosets $g_w(K_x \cap K_y)$ are all distinct. The claim follows because $\mathfrak{B}_n$ has $(q+1)q^{n-1}$ elements.

Before analyzing in more detail the group Aut($\mathfrak{X}$) for a general homogeneous tree, we give here a description of the simplest nontrivial case, that for which q+1=2 (the reader may verify that a homogeneous tree of degree 1 consists of just one edge). If q+1=2, then $\mathfrak{X}$ consists of a doubly infinite chain, that is a geodesic $\{\ldots x_{-2},x_{-1},x_0,x_1,\ldots\}=\mathfrak{X}$, with $\mathbf{d}(x_i,x_{i+1})=1$. We may therefore identify $\mathfrak{X}$ with the integers through the correspondence $n \to x_n$. It is clear that translations on $\mathfrak{X}$ form a group generated by a step-1 translation and isomorphic to $\mathbb{Z}$. For every $x_n \in \mathfrak{X}$, there is only one nontrivial rotation which fixes x_n and maps x_{n+j} to x_{n-j}, for $j \in \mathbb{Z}$. If k_0 is the rotation about x_0 and τ is a step-1 translation, $\tau^n k_0 \tau^{-n}$ is a rotation which fixes x_n. All rotations of Aut($\mathfrak{X}$) are of this form. Similarly there is only one inversion g_0 on the edge $\{x_0,x_1\}$ and all the inversions of Aut($\mathfrak{X}$) are of the form $\tau^n g_0 \tau^{-n}$, $n \in \mathbb{Z}$. It is easy to see that Aut($\mathfrak{X}$) can be written as the semidirect product of the group generated by a step-1 translation τ, and the two-element group generated by k_0. The translations of even step form a subgroup of Aut($\mathfrak{X}$), still isomorphic to $\mathbb{Z}$. This subgroup has two orbits. Every edge intersects both orbits, and therefore the group generated by the translations of even step and one inversion acts transitively on $\mathfrak{X}$. It is easy to see

(see (6.4) below, for a more general construction) that this group is generated by two inversions a and b on edges having one vertex in common. Observe that $a^2=b^2=e$ (the identity) and that the group generated by a and b is isomorphic to the free product $\mathbb{Z}_2*\mathbb{Z}_2$ of two copies of $\mathbb{Z}_2=\mathbb{Z}/2\mathbb{Z}$. A step-1 translation may be obtained as the product of an inversion followed by a rotation. This means that $Aut(\mathfrak{X})$ is also the semidirect product of the group generated by two inversions on adjacent edges and the two-element group generated by the rotation about their common vertex. In other words $Aut(\mathfrak{X})$ is isomorphic to $(\mathbb{Z}_2*\mathbb{Z}_2)\rtimes\mathbb{Z}_2$.

5. Compact maximal subgroups. Besides the stabilizer of a vertex K_x, we can also consider the stabilizer of an edge $K_{\{a,b\}}=\{g\in Aut(\mathfrak{X}):\ g(\{a,b\})=\{a,b\}\}$. This group is also compact and indeed $K_{\{a,b\}}=(K_a\cap K_b)\cup g(K_a\cap K_b)$ where g is an inversion with g(a)=b and g(b)=a. The subgroups $K_{\{a,b\}}$ are also all conjugate to each other, and open. We will show next that the groups K_x and $K_{\{a,b\}}$ are the only maximal compact subgroups of $Aut(\mathfrak{X})$. We first need an elementary observation.

(5.1) *PROPOSITION. Let* G *be a subgroup of* $Aut(\mathfrak{X})$; *then the closure of* G *is compact if and only if, for every* $x\in\mathfrak{X}$, *the orbit* $G(x)=\{g(x):\ g\in G\}$ *is finite.*

PROOF. If G is compact, then G(x) is compact in the discrete metric space $\mathfrak{X}$, and therefore G(x) is finite. Conversely, suppose that, for some $x\in\mathfrak{X}$, $G(x)=\{g_1(x),\dots,g_n(x)\}$ is finite. Then $G\subseteq\bigcup_{j=1}^{n}\{g:\ g(x)=g_j(x)\}$. But $\{g:\ g(x)=g_j(x)\}=\{g_jg:\ g(x)=x\}=g_jK_x$, which is compact.∎

(5.2) *THEOREM. Every compact subgroup of* $Aut(\mathfrak{X})$ *is contained in a group of the type* K_x, *for some* $x\in\mathfrak{X}$, *or a group of the type* $K_{\{a,b\}}$, *for some edge* $\{a,b\}\in\mathfrak{E}$. *These groups are maximal subgroups of* $Aut(\mathfrak{X})$.

PROOF. Let G be a compact subgroup of $\mathrm{Aut}(\mathfrak{X})$. Since for each $x\in\mathfrak{X}$ the orbit $G(x)$ is finite, so is the subtree of $\mathfrak{X}$ generated by $G(x)$. This subtree is mapped by G into itself. Therefore there exists a minimal G-invariant subtree $\mathfrak{Y}\subseteq\mathfrak{X}$. If $\mathfrak{Y}$ contains more than two vertices, then it contains vertices of degree 1 and vertices of degree greater than 1. Since the degree is invariant under G the subtree of $\mathfrak{Y}$ obtained by omitting the vertices of degree 1 is still invariant, contradicting the minimality of $\mathfrak{Y}$. Therefore $\mathfrak{Y}$ consists of one vertex or of one edge, which implies that G is contained in some K_x or in some $K_{\{a,b\}}$. It remains to show that K_x and $K_{\{a,b\}}$ are maximal compact subgroups. Observe that (3.4) implies that for no vertex $x\in\mathfrak{X}$, and no edge $\{a,b\}\in\mathfrak{E}$, can we have $K_x\subseteq K_{\{a,b\}}$ or $K_{\{a,b\}}\subseteq K_x$. This shows, by the first part of the theorem, that, if G is compact and $K_x\subseteq G$, then, for some y, $K_x\subseteq G\subseteq K_y$, which implies $x=y$ and $G=K_x$; and, if G is compact and $K_{\{a,b\}}\subseteq G$, then $K_{\{a,b\}}\subseteq G\subseteq K_{\{c,d\}}$, for some $\{c,d\}\in\mathfrak{E}$, which implies $\{a,b\}=\{c,d\}$ and $G=K_{\{a,b\}}$.∎

We conclude with a result which will be used in the next section.

(5.3) *LEMMA. Let* G *be a subgroup of* $\mathrm{Aut}(\mathfrak{X})$ *and* x *a vertex. If the orbit* $G(x)$ *contains three distinct vertices on the same geodesic, then* G *contains a translation.*

PROOF. If $x,y,t\in G(x)$ are in the same chain, we may assume that the chain $[x,y]$ joining x to y contains t, that is $[x,y] = [x,t]\cup[t,y]$, and $[x,t]\cap[t,y]=\{t\}$. Let $g,h\in G$ be such that $g(t)=x$ and $h(t)=y$. Suppose $[t,x]=\{t,x_1,x_2,\ldots,x_n=x\}$ and $[t,y]=\{t,y_1,y_2,\ldots,y_m=y\}$. By (3.1), if $g(x_1)\neq x_{n-1}$, then g is a translation. Similarly, if $h(y_1)\neq y_{m-1}$, then h is a translation and the lemma is proved. Therefore we can suppose that $g(x_1)=x_{n-1}$ and $h(y_1)=y_{m-1}$. But $x_1\neq y_1$ implies $hg^{-1}(x_{n-1}) =$

$h(x_1) \neq h(y_1) = y_{m-1}$. Therefore (3.1) implies that hg^{-1} is a translation because $hg^{-1}(x)=y$. ∎

6. Discrete subgroups. We will now discuss a class of *discrete* subgroups of $\mathrm{Aut}(\mathfrak{X})$, which have trivial intersections with every K_x.

(6.1) *DEFINITION. We say that a subgroup* Γ *of* $\mathrm{Aut}(\mathfrak{X})$ *acts faithfully and transitively on* $\mathfrak{X}$*, or that it is a faithful transitive subgroup, if, for every* $x,y \in \mathfrak{X}$*, there exists* $g \in \Gamma$ *such that* $g(x)=y$*, and if* $\Gamma \cap K_x$ *is the identity for each* $x \in \mathfrak{X}$*.*

If Γ is a faithful transitive group of automorphisms, then, given any vertex $\mathbf{o} \in \mathfrak{X}$, the map $g \to g\mathbf{o}$ is bijective. Furthermore Γ is discrete in $\mathrm{Aut}(\mathfrak{X})$. There is an easy characterization of the faithful transitive subgroups of $\mathrm{Aut}(\mathfrak{X})$ for which we need first the following technical result.

(6.2) *LEMMA. Suppose that there exist a vertex* $\mathbf{o} \in \mathfrak{X}$ *and a subset* A *of* $q+1$ *elements of* $\mathrm{Aut}(\mathfrak{X})$*, such that*

(a) $A = A^{-1}$

(b) $A(\mathbf{o}) = \{a(\mathbf{o}) : a \in A\} = \{y : \mathbf{d}(\mathbf{o},y)=1\}$.

Then

(1) *for every finite sequence* $a_1,\dots,a_n$ *of elements of* A, $\mathbf{d}(\mathbf{o}, a_1 \dots a_n(\mathbf{o})) \le n$,

(2) *for every* $x \in \mathfrak{X}$*, with* $\mathbf{d}(\mathbf{o},x)=n$*, there exists a finite sequence* $a_1,\dots,a_n$ *of elements of* A*, such that* $x = a_1 a_2 \dots a_n(\mathbf{o})$ *and* $a_i a_{i+1} \neq 1$*, for* $i=1,\dots,n-1$.

PROOF. To prove (1) we use induction on n. The assertion is true for $n=1$. If $a_1,\dots,a_{n+1} \in A$, then

$$\mathbf{d}(\mathbf{o}, a_1 \dots a_{n+1}(\mathbf{o})) =$$
$$\mathbf{d}(a_1^{-1}(\mathbf{o}), a_2 \dots a_{n+1}(\mathbf{o})) \le \mathbf{d}(\mathbf{o}, a_2 \dots a_{n+1}(\mathbf{o})) + 1 = n+1,$$

because $\mathbf{d}(a_1^{-1}(\mathbf{o}),\mathbf{o})=1$. To prove (2) we observe again that the assertion is true for $n=1$. Supposing that it is true for $n \ge 1$, let $x\in\mathfrak{X}$, and $\mathbf{d}(x,\mathbf{o})=n+1$. Let $\{\mathbf{o}=x_0,x_1,\ldots,x_{n+1}=x\}$ be the chain $[\mathbf{o},x]$. There exists only one element $a\in A$ such that $a(\mathbf{o})=x_1$. But $\mathbf{d}(a^{-1}(x),\mathbf{o})=\mathbf{d}(x,x_1)=n$, and therefore $a^{-1}(x)=a_1a_2\ldots a_n(\mathbf{o})$, with $a_ia_{i+1}\neq1$, for $i=1,2,\ldots,n-1$. It follows that $x=aa_1a_2\ldots a_n(\mathbf{o})$. It remains to prove that $aa_1\neq1$. But $aa_1=1$ implies $\mathbf{d}(\mathbf{o},x)\le n-1$, by part (1). Since $\mathbf{d}(\mathbf{o},x)=n$, we must have $aa_1\neq1$. ∎

(6.3) *THEOREM. Let* A *be a subset of* Aut($\mathfrak{X}$) *which satisfies the hypothesis of* (6.2). *Then* A *generates a faithful transitive subgroup* Γ *of* Aut($\mathfrak{X}$). *Furthermore* Γ *is isomorphic to the free product of* t *copies of* $\mathbb{Z}$ *and* s *copies of the two-element group* $\mathbb{Z}_2$, *for some* $t\ge0$ *and* $s\ge0$ *satisfying* $2t+s=q+1$. *Finally every faithful transitive subgroup* Γ *of* Aut($\mathfrak{X}$) *is generated by a subset* A *satisfying the hypothesis of* (6.2).

PROOF. Let $\mathbf{o}$ be a fixed vertex of $\mathfrak{X}$, then the map $a\to a(\mathbf{o})$ is bijective from A onto $\mathfrak{B}_1=\{x\colon \mathbf{d}(x,\mathbf{o})=1\}$. If $a\in A$, then either $a^2(\mathbf{o})=\mathbf{o}$, and therefore a is an inversion which leaves the edge $\{\mathbf{o},a(\mathbf{o})\}$ fixed, or $a(\mathbf{o})\neq\mathbf{o}$, and by (3.1) a cannot be other than a step-1 translation. If a is an inversion then $a(\mathbf{o})=a^{-1}(\mathbf{o})$, and since $A=A^{-1}$, it must be $a=a^{-1}$. Let $a_1,\ldots a_s$ be the inversions of period 2 contained in A. Then $A=\{a_1,\ldots,a_s,\tilde a_1,\ldots,\tilde a_t,\tilde a^{-1},\ldots,\tilde a_t^{-1}\}$, where $q+1=s+2t$, and $\tilde a_1,\ldots,\tilde a_t$ are step-1 translations. Let $G=\mathbb{Z}*\ldots*\mathbb{Z}*\mathbb{Z}_2*\ldots*\mathbb{Z}_2$ be the free product of t copies of $\mathbb{Z}$ and s copies of $\mathbb{Z}_2$. Then Γ is the image of G under the canonical homomorphism which assigns the generators of order two of G to the elements $a_1,\ldots,a_s$, and the generators of infinite order of G to $\tilde a_1,\ldots,\tilde a_t$. Observe that the set G_n of elements of G which have length no greater than n has the same number of elemets as $\mathfrak{X}_n=\{x\in\mathfrak{X}\colon \mathbf{d}(\mathbf{o},x)\le n\}$. Let Γ_n be the image of G_n under the homomorphism above, that is let Γ_n be the set of elements of Γ which can be written as words of length no greater than n in the elements of A. Then the map $g\to g(\mathbf{o})$ maps Γ_n surjectively onto $\mathfrak{X}_n$. Therefore Γ_n has at least the same cardinality as $\mathfrak{X}_n$ or G_n. This shows that the canonical homomorphism is injective and that Γ is isomorphic to G. It also shows that the map $g \to g(\mathbf{o})$ is injective and therefore Γ is a faithful subgroup. Conversely, let Γ be a faithful transitive subgroup of $\mathrm{Aut}(\mathfrak{X})$, and let $\mathbf{o}$ be a fixed vertex of $\mathfrak{X}$. Let $A=\{g\in\Gamma:\ \mathbf{d}(\mathbf{o},g(\mathbf{o}))=1\}$. Since $\mathbf{d}(g(\mathbf{o}),\mathbf{o})=\mathbf{d}(g^{-1}(\mathbf{o}),\mathbf{o})$ we have that $A=A^{-1}$. Since $g\to g(\mathbf{o})$ is a bijection of Γ onto $\mathfrak{X}$, it follows that A is in one-to-one correspondence with $\mathfrak{B}_1=\{x:\ \mathbf{d}(x,\mathbf{o})=1\}$. Therefore A satisfies the hypothesis of (6.3) and generates a faithful transitive subgroup of Γ. Clearly this subgroup must be all of Γ. ▮

We have shown in particular that, given $\mathbf{o}\in\mathfrak{X}$, any faithful transitive subgroup Γ is generated by the set $\{g\in\Gamma:\ \mathbf{d}(g(\mathbf{o}),\mathbf{o})=1\}$. Observe that $\mathrm{Aut}(\mathfrak{X})$ can be written as the product of the stabilizer K_o of the vertex $\mathbf{o}$, and any faithful transitive subgroup Γ. Indeed if $g\in\mathrm{Aut}(\mathfrak{X})$, there exists $g'\in\Gamma$ such that $g'(\mathbf{o})=g(\mathbf{o})$, which implies $g^{-1}g'\in K_o$ and $g=kg'$, for some $k\in K_o$; clearly $K_o\cap\Gamma=\{e\}$.

A consequence of the decomposition $\mathrm{Aut}(\mathfrak{X})=K_o\Gamma$ is that $\mathrm{Aut}(\mathfrak{X})$ is unimodular and its Haar measure is the product of the Haar measure on K_o (which may be normalized so as to have total mass 1) and the Haar measure on Γ (which may be normalized so that points have mass 1). A similar decomposition is true for every closed subgroup of $\mathrm{Aut}(\mathfrak{X})$ which contains a subset A satisfying the hypothesis of (6.3).

7. Cayley graphs which are trees. In Section 2, Fig.5, we described the tree associated to a free group with two generators. Indeed to every homogeneous tree we can associate a finitely generated group, in such a way that the elements of the group are identified with the vertices of the tree and the tree is the *Cayley graph* of the group with respect to a given set of generators. We recall that, if Γ is a group and E is a set of generators of Γ, the Cayley graph of Γ is a graph with vertices Γ and edges $\{\{x,y\}: x,y\in\Gamma$, there exists $s\in E$, with $xs=y\}$. We observe that the Cayley graph as defined above is always connected and that left multiplication by an element of Γ is an automorphism of the Cayley graph of Γ (it maps edges into edges). Furthermore Γ acts faithfully and transitively on its graph. Therefore (6.3) tells us that the graph of Γ is a locally finite homogeneous tree if and only if Γ is the free product of t copies of $\mathbb{Z}$ and s copies of $\mathbb{Z}_2$, with $2s+t=q+1$. Figs 9-12 illustrate the graphs of $\mathbb{Z}$, $\mathbb{Z}_2*\mathbb{Z}_2$, $\mathbb{Z}*\mathbb{Z}_2$, and $\mathbb{Z}_2*\mathbb{Z}_2*\mathbb{Z}_2$. The graph of $\mathbb{Z}*\mathbb{Z}$ was given in Section 2, Fig.5.

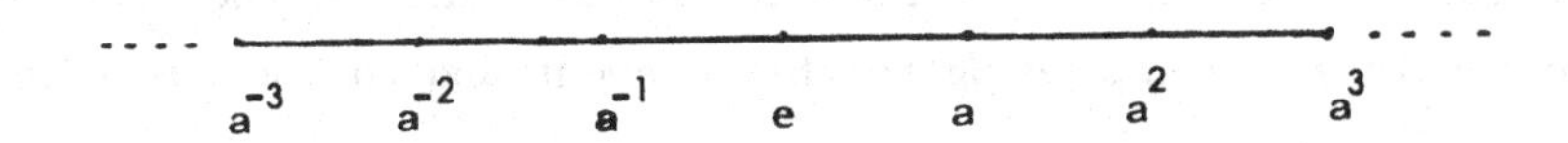

Fig.9 $\Gamma = \mathbb{Z}$, a is a generator of infinite order

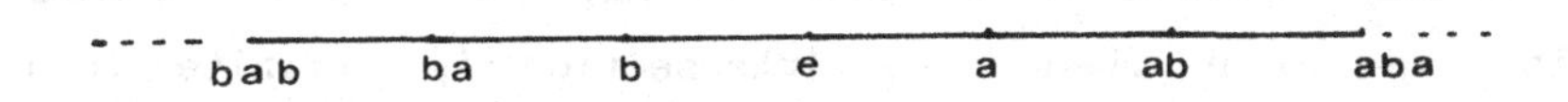

Fig.10 $\Gamma = \mathbb{Z}_2*\mathbb{Z}_2$, a and b are generators of order 2

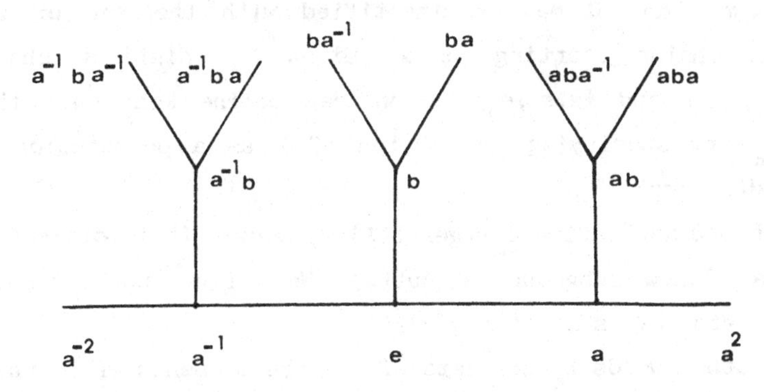

Fig.11 $\Gamma = \mathbb{Z}*\mathbb{Z}_2$, a is of infinite order and b is of order 2

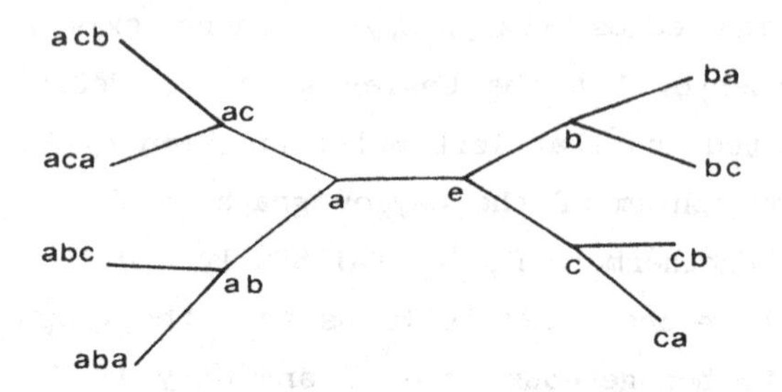

Fig.12 $\Gamma = \mathbb{Z}_2 * \mathbb{Z}_2 * \mathbb{Z}_2$, a, b and c have order 2

We observe that, if the graph of Γ is a tree $\mathfrak{X}$, then the length of an element of Γ as a reduced word in the generators and their inverses is nothing but the distance in the tree of the vertices corresponding to that element and to the identity.

8. Amenable subgroups. In order to describe other notable subgroups of $\mathrm{Aut}(\mathfrak{X})$, we need to consider the action of $\mathrm{Aut}(\mathfrak{X})$ on the boundary Ω of $\mathfrak{X}$. We recall that Ω is the set of equivalence classes of infinite chains, two infinite chains being equivalent when their intersection is infinite (and therefore cofinite in each of the two chains). An automorphism $g \in \mathrm{Aut}(\mathfrak{X})$ maps an infinite chain into an infinite chain, preserving the equivalence classes. Therefore $\mathrm{Aut}(\mathfrak{X})$ acts naturally on Ω. We observe that, if $x \in \mathfrak{X}$, the group K_x acts transitively on Ω. Each element $\omega \in \Omega$ has a representative of the type $[x,\omega)$, and Ω may be identified with the set of all infinite chains starting at x. Given two distinct chains $\{x, t_1, t_2, \ldots\}$ and $\{x, s_1, s_2, \ldots\}$ we can define $k \in K_x$ such that $k(s_n) = t_n$, by specifying the action of k as a permutation of $\mathfrak{B}_n = \{y\colon \mathbf{d}(x,y) = n\}$.

Let $\omega \in \Omega$ and define $G_\omega = \{g \in \mathrm{Aut}(\mathfrak{X})\colon g(\omega) = \omega\}$. It is clear that G_ω is a closed subgroup of $\mathrm{Aut}(\mathfrak{X})$. We define now $B_\omega = \{g \in G_\omega\colon$ there exists $x \in \mathfrak{X}$, such that $g(x) = x\}$.

In other words B_ω consists of all the elements of G_ω which are *rotations* about some vertex. It is not difficult to see that B_ω is a group. Indeed, given $x,y\in\mathfrak{X}$, the infinite chains $[x,\omega)$ and $[y,\omega)$ have infinite intersection. If g and g' are elements of B_ω and $g(x)=x$, $g'(y)=y$, for every $t\in[x,\omega)\cap[y,\omega)$, we have $g'g^{-1}\in G_\omega\cap K_t\subseteq B_\omega$. The group B_ω is the union of compact subgroups $B_\omega=\bigcup_{x\in\mathfrak{X}} K_x\cap G_\omega$, and indeed, if $\{x_0,x_1,x_2,...\}$ is any infinite chain belonging to the equivalence class ω, then $B_\omega=\bigcup_{n=0}^{\infty} K_{x_n}\cap G_\omega$. By definition $K_x\cap G_\omega$ is open in G_ω and in B_ω, for each x. Therefore B_ω is the union of countably many compact open subgroups, which implies that B_ω is open in G_ω and is amenable. We will show now that B_ω is normal in G_ω. Let $g\in G_\omega$ and suppose that g is not a rotation. Since an inversion cannot fix any point of Ω, g cannot be an inversion. Therefore we may assume that g is a translation along a geodesic $\gamma=(\omega,\omega')=\{...,x_{-1},x_0, x_1,...\}$. But $B_\omega=\bigcup_n G_\omega\cap K_{x_n}$, and, if $k\in G_\omega\cap K_{x_n}$, then $gkg^{-1}\in G_\omega\cap K_{g(x_n)}$. Therefore B_ω is normal in G_ω. Let now $\gamma=(\omega,\omega')$ be a fixed geodesic and let $M_\gamma=\bigcap_{x\in\gamma} K_x$. Then M_γ is a compact group contained in B_ω. Observe that if τ is a fixed step-1 translation on γ, moving γ in the direction of ω, and g any step-n translation on γ, then $\tau^n g^{-1}\in M_\gamma$. Thus every translation on γ may be written as the product of an element of M_γ and an element of the group generated by τ (which of course is isomorphic to $\mathbb{Z}$). If $g\in G_\omega$, and $g\notin B_\omega$, then g is a translation on a geodesic $\gamma'=(\omega'',\omega)$, where ω'' may be different from ω'. Replacing g with g^{-1} if necessary, we may assume that g moves γ' towards ω. Let $x\in\gamma\cap\gamma'$ and suppose that $\mathbf{d}(x,g(x))=n$. Then $\tau^{-n}g(x)=x$. Therefore $\tau^{-n}g\in B_\omega$. We have proved therefore that every element of G_ω may be written as the product of an element of B_ω and an element of the group generated by a step-1 translation on the geodesic $\gamma=(\omega',\omega)$. The latter group is abelian, and isomorphic to $\mathbb{Z}$. In conclusion we have shown that G_ω is the semidirect product of a copy of $\mathbb{Z}$ and the group B_ω. In particular G_ω is amenable.

We saw that any two translations along the same geodesic γ differ by an element of M_γ. The quotient B_ω/M_γ does not correspond to any subgroup of B_ω, but, as will be seen later, may be naturally identified with the orbits of B_ω, which are called the *horocycles*, relative to ω.

We should also observe that the set of all automorphisms which are *translations* along the geodesic $\gamma=(\omega',\omega)$, or are the identity on this geodesic, is a group containing M_γ as a normal subgroup. Thus M_γ is the kernel of a homomorphism of this group onto $\mathbb{Z}$.

Given the geodesic $\gamma=(\omega',\omega)$ we may consider the group $G_\gamma=\{g\in \mathrm{Aut}(\mathfrak{X}): g(\gamma)=\gamma\}$. If κ is either a rotation or an inversion interchanging ω and ω', then $G_\gamma=(G_\omega\cap G_{\omega'})\cup \kappa(G_\omega\cap G_{\omega'})$. It follows that the closed amenable subgroup $G_\omega\cap G_{\omega'}$ has index 2 in G_γ, and therefore G_γ is amenable. We will now prove that the only amenable closed subgroups of $\mathrm{Aut}(\mathfrak{X})$ are the closed subgroups of the groups K_x, $K_{\{a,b\}}$, G_ω, and G_γ.

(8.1) *THEOREM. Let* G *be a subgroup of* $\mathrm{Aut}(\mathfrak{X})$*, and suppose that* G *contains no translations. Then* $G\subseteq K_x$ *for some vertex* x*, or* $G\subseteq K_{\{a,b\}}$ *for some edge* $\{a,b\}$*, or else* $G\subseteq B_\omega$ *for some* $\omega\in\Omega$*.*

PROOF. By (5.3) we know that no orbit of G contains three vertices on the same geodesic. If G has compact closure, it is contained in a maximal compact subgroup which, by (5.2), is either a K_x or a $K_{\{a,b\}}$. It suffices to show, by (5.1) that, if $G(x)$ is infinite for some vertex x, then $G\subseteq G_\omega$ for some $\omega\in\Omega$. Indeed if G has no translations and $G\subseteq G_\omega$, then G consists only of rotations, since no inversion can fix a point of the boundary. This implies that $G\subseteq B_\omega$. Suppose now that one, and therefore every, orbit of G is infinite. Then every orbit has a limit point in the compact space $\mathfrak{X}\cup\Omega$. We want to show that, for some vertex x, $G(x)$ has only one limit point. Since the closure of $G(x)$ in $\mathfrak{X}\cup\Omega$ is G-invariant, this will show that there is a point $\omega\in\Omega$, which is fixed by G, that is that $G\subseteq G_\omega$. We shall reason by contradiction observing that, if ω and ω' are two distinct limit points of $G(x)$, then $G(x)\cap(\omega,\omega')=\emptyset$. Indeed let t belong to the intersection of (ω,ω') and $G(x)$, and let y, z be vertices of (ω,ω') at distance 1 from t and in the direction of ω and ω', respectively. Then $\mathfrak{C}(t,z)$ and $\mathfrak{C}(t,y)$ are disjoint neighborhoods of ω and ω' respectively. Therefore there exist vertices $y'\in\mathfrak{C}(t,y)\cap G(x)$ and $z'\in\mathfrak{C}(t,z)\cap G(x)$. The chain $[y',z']$ includes the point t and therefore, by (5.3), G contains a translation. Let now $G(x_n)$ be the denumerable set of infinite orbits into which $\mathfrak{X}$ is partitioned. We are assuming the contrary of what we want to show, that each of the orbits has more than one limit point in Ω. Let Ω_1 be the set of limit points of $G(x_1)$, and let $\mathfrak{Y}_1$ be the subtree consisting of all the chains (ω,ω') with $\omega,\omega'\in\Omega_1$. Observe that $\mathfrak{Y}_1$ is G-invariant, and $\mathfrak{Y}_1\cap G(x_1)=\emptyset$. Therefore we can construct inductively a sequence of subtrees $\mathfrak{Y}_k$, and a sequence of vertices y_k, such that $\mathfrak{Y}_k\cap G(y_k)=\emptyset$, $y_k\in\mathfrak{Y}_{k-1}$, and $\bigcap\mathfrak{Y}_k=\emptyset$. It suffices indeed to let $x_1=y_1$, and $y_k=x_{n_k}$, where n_k is the first index such that $x_{n_k}\in\mathfrak{Y}_{k-1}$ and finally, if Ω_k is the set of limit points of $G(y_k)$, we let $\mathfrak{Y}_k$ to be the tree consisting of all infinite geodesics (ω,ω') with $\omega,\omega'\in\Omega_k$. Clearly $\mathfrak{Y}_k\cap G(x_n)=\emptyset$, if k is sufficiently large. Furthermore $G(y_k)\subseteq\mathfrak{Y}_{k-1}$, which implies $\mathfrak{Y}_k\subseteq\mathfrak{Y}_{k-1}$ and $\Omega_k\subseteq\Omega_{k-1}$. This means that the intersection of the closed nested sets Ω_k is nonempty. This intersection is an invariant subset of Ω which cannot consist of more than one point because the intersection of the $\mathfrak{Y}_k$ is empty. Thus there exists a point of Ω which is invariant under G and G is contained in G_ω. ▮

(8.2) *LEMMA. Let γ and γ' be two geodesics having at most one vertex in common. Let τ and τ' be translations on γ and γ' respectively. Then the subgroup generated by τ and τ' is discrete and isomorphic to a free group.*

PROOF. Let $x\in\gamma$ be the point of minimal distance from γ'. By hypothesis x is uniquely defined, because either $\gamma\cap\gamma'=\emptyset$, or $\gamma\cap\gamma'=\{x\}$. Let C'(x) be the union of all infinite chains starting at x and containing no other point of γ. Let C(x) be the complement of C'(x) together with the vertex x. Then $C'(x)\cap C(x)=\{x\}$, and $\gamma'\subseteq C'(x)$. Furthermore any chain connecting a point of C'(x) with a point of C(x) goes through x. Observe that, if γ'' is a geodesic contained in C'(x) (resp. C(x)), then $\tau(\gamma'')$ (resp. $\tau'(\gamma'')$) is contained in C(x) (resp. C'(x)) and does not intersect γ (resp. γ'). Let now w be a reduced nonempty word in τ and τ' which contains the letter τ' at least once. We observe that $w(\gamma)$ is disjoint from γ. Indeed w is the product of powers of τ and τ'; a power of τ maps γ onto itself; therefore we may assume that w has a τ' as its last letter. A nonzero power of τ' maps γ onto a geodesic contained in C'(x) which does not intersect γ. Therefore the first application of a power of τ' and all successive applications of powers of τ and τ' map γ into geodesics which do not meet γ. We have proved that $w(\gamma)\cap\gamma=\emptyset$, if w contains a power of τ'. On the other hand if w is a nonzero power of τ, then w(x) is not the identity on γ. We have thus proved that any reduced word in τ and τ' is not the identity as an element of $\mathrm{Aut}(\mathfrak{X})$, and therefore that τ and τ' generate a free group. In addition the intersection of this group with any stabilizer of a point of γ is the identity, and therefore this free group is discrete. ∎

We shall characterize now the maximal closed amenable subgroups of $\mathrm{Aut}(\mathfrak{X})$.

(8.3) *THEOREM. Let* G *be a closed subgroup of* $\mathrm{Aut}(\mathfrak{X})$. *Then* G *is amenable if and only if one of the following occurs: (i)* G *is compact; (ii)* $G\subseteq G_\omega$ *for some* $\omega\in\Omega$; *(iii)* $G\subseteq G_\gamma$ *for some geodesic* γ.

PROOF. We have observed already that the groups G_ω and G_γ are amenable. Compact groups are of course amenable. Conversely, suppose that G is amenable. If G has no translation, then by

(8.1) it is compact, or it is contained in B_ω. We may assume therefore that G has a nontrivial translation τ on a geodesic γ. For the rest of the proof the translation τ and the geodesic γ will be fixed. We observe first that, if $g\in G$, then $\gamma\cap g(\gamma)$ is infinite. Indeed, $g\tau g^{-1}$ is a nontrivial translation on the geodesic $g(\gamma)$. But G is amenable, and therefore it cannot contain a discrete free group with two generators. This means, by (8.2), that $g(\gamma)\cap\gamma\neq\emptyset$. Furthermore $g(\gamma)\cap\gamma$ cannot be finite, because otherwise, for some integer $h\neq 0$, $\tau^h[g(\gamma)\cap\gamma]\cap g(\gamma)=\emptyset$, which implies $g(\gamma)\cap\tau^h g(\gamma)=\emptyset$, which is again impossible, by (8.2), because $\tau^h g\tau g^{-1}\tau^{-h}$ is a translation belonging to G on the geodesic $\tau^h g(\gamma)$. We have proved that $g(\gamma)\cap\gamma$ is infinite for every $g\in G$. On the other hand if h is another element of G, we also have that $h(\gamma)\cap g(\gamma)$ is infinite because $h\tau h^{-1}$, and $g\tau g^{-1}$ are translations on $h(\gamma)$ and $g(\gamma)$, respectively. We shall prove now that either $G\subseteq G_\gamma$ or G stabilizes one of the end points of γ. Let $\gamma=(\omega,\omega')$, and $E=\{\omega,\omega'\}$. If $g(\gamma)=\gamma$ then $g(E)=E$, but if $g(\gamma)\neq\gamma$ then $g(\gamma)\cap\gamma$ is an infinite chain and $g(E)\cap E$ consists of only one point. If for every g $g(\gamma)=\gamma$, then $G\subseteq G_\gamma$. Assume now that there exists an element g of G such that $g(\gamma)\neq\gamma$, and let ω be the point in $g(E)\cap E$. We shall prove that for every $h\in G$, such that $h(\gamma)\neq\gamma$, we have $h(E)\cap E=\{\omega\}$. In other words we shall prove that $h(\gamma)\cap g(\gamma)\cap\gamma$ is an infinite chain. This is clear because, if $h(\gamma)\cap g(\gamma)\cap\gamma$ is finite, then either $h(\gamma)\cap g(\gamma)$ is finite, or $h(\gamma)\cap g(\gamma)\cap\gamma$ consists of only one point, and therefore $h(\gamma)\cap\tau g(\gamma)$ is empty. Both these things are impossible and therefore $\bigcap_{g\in G} g(E)=\{\omega\}$ is a G-invariant subset of Ω, which means that $G\subseteq G_\omega$. ∎

We can deduce from the results of this section that *a nonamenable closed subgroup of* Aut($\mathfrak{X}$) *necessarily contains a free discrete group in two generators.* Indeed if G is not amenable, it must contain a translation τ on a geodesic γ (8.3). But the proof of (8.3) shows that if $g(\gamma)\cap\gamma\neq\emptyset$, for every $g\in G$, then $G\subseteq G_\gamma$ or $G\subseteq G_\omega$ for some ω. Therefore if G is not amenable, for some $g\in G$, $g(\gamma)\cap\gamma=\emptyset$, which implies, by (8.3), that τ and $g\tau g^{-1}$ generate a free discrete subgroup.

We should also remark that G_ω acts transitively on $\mathfrak{X}$ and on $\Omega\setminus\{\omega\}$. It is clear that the stabilizer of a vertex or of an edge or of a geodesic does not act transitively on $\mathfrak{X}$. In other words we may say that *a closed subgroup of* Aut($\mathfrak{X}$), *acting transitively on* $\mathfrak{X}$, *is amenable if and only if it stabilizes a point of* Ω.

9. Orbits of amenable subgroups. In this section we shall describe the orbits on the set of vertices $\mathfrak{X}$ and on the boundary Ω of the subgroups of Aut($\mathfrak{X}$) which were described in the previous sections.

If $x\in\mathfrak{X}$ the orbits of K_x are exactly the sets $\mathfrak{B}_n=\{y: \mathbf{d}(x,y)=n\}$. Furthermore, a closed subgroup of K_x acts transitively on Ω if and only if it acts transitively on $\mathfrak{B}_n$ for every n. This follows immediately from the fact that Ω can be described as the set of infinite chains starting at x. If $\{a,b\}$ is an edge, then $K_{\{a,b\}}\cap K_a$ acts transitively on the chains which start at a and pass through b. Since $K_{\{a,b\}}$ contains an inversion exchanging a and b, it follows that $K_{\{a,b\}}$ acts transitively on Ω. The orbits of $K_{\{a,b\}}$ on $\mathfrak{X}$ are $\{a,b\}$, the sets $\{y: \mathbf{d}(a,y)=1,\ y\neq b\}\cup\{y: \mathbf{d}(y,b)=1,\ y\neq a\}$, and in general, for $n>1$, $\{y: \mathbf{d}(y,a)=n,\ b\notin[a,y]\}\cup\{y: \mathbf{d}(y,b)=n,\ a\notin[b,y]\}$.

Let γ be a doubly infinite geodesic. The group G_γ has γ as an orbit and the other orbits are the sets $\{y: \mathbf{d}(y,\gamma)=n\}$. On the boundary Ω, G_γ has two orbits: $\{\omega,\omega'\}$ and $\Omega\setminus\{\omega,\omega'\}$, where ω and ω' are the two ends of γ.

For ω a point of the boundary, we have that B_ω, and *a fortiori* G_ω, act transitively on $\Omega\setminus\{\omega\}$. In addition G_ω acts transitively on $\mathfrak{X}$. This is proved as follows. Given $x,y\in\mathfrak{X}$, the chains $[x,\omega)$ and $[y,\omega)$ intersect in the chain $[z,\omega)$. Suppose $\mathbf{d}(y,z)\geq\mathbf{d}(x,z)$, then there is a rotation k, which fixes z and ω and maps x into a point of $[y,z]$. Thus $k(x)\in[y,\omega]$. Let γ be any geodesic containing $[y,\omega)$, and let τ be a translation along γ and such that $\tau(k(x))=y$. Then $\tau\in G_\omega$ and therefore $\tau k\in G_\omega$.

The group B_ω does not act transitively on $\mathfrak{X}$. Indeed, if x and y are different elements lying on the same geodesic having ω as one of the ends, no element of B_ω can map x into y. To describe the orbits of B_ω we introduce an equivalence relation on $\mathfrak{X}$ as follows. Let $x,y\in\mathfrak{X}$ and suppose that $[x,\omega)\cap[y,\omega)=[z,\omega)$. Then we say that x is equivalent to y if $\mathbf{d}(z,x)=\mathbf{d}(z,y)$. The equivalence classes of this relation are called *horocycles* of ω.

The *horocycles* are exactly the orbits of B_ω. Let $\gamma=(\omega',\omega)=\{\ldots,s_{-1},s_0,s_1,\ldots\}$, that is assume $\lim s_n=\omega$. Then $H_n=\{gs_n:\ g\in B_\omega\}$ is a horocycle and $\bigcup H_n=\mathfrak{X}$. Each horocycle H_n divides $\mathfrak{X}$ into two subsets, intersecting in the horocycle H_n: the subtree "inside" the horocycle, $\bigcup_{k\geq n} H_k$, and the subset "outside" the horocycle, $\bigcup_{k\leq n} H_k$. Observe that every geodesic having ω as one of the end points intersects each horocycle exactly once. Fig. 13 gives a geometric picture of the horocycles, for the case q=2.

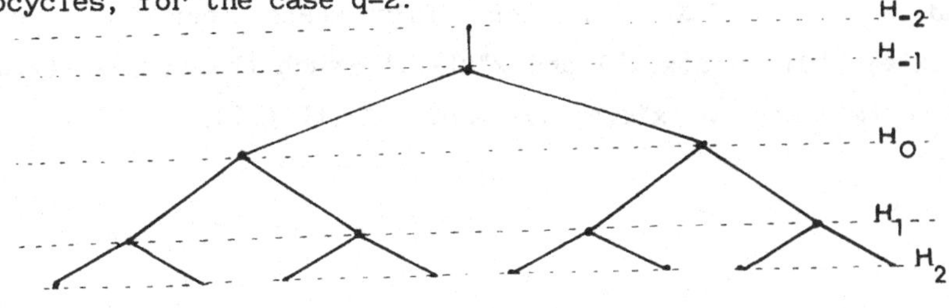

Fig.13

This picture makes it clear why horocycles are sometimes said to represent the *generations* with respect to a common *mythical ancestor* ω.

10. Groups with transitive action on the boundary. We have seen that compact maximal subgroups of $\mathrm{Aut}(\mathfrak{X})$ act transitively on the boundary. The following proposition deals with noncompact groups which act transitively on Ω.

(10.1) *PROPOSITION. Let* G *be a closed subgroup of* $\mathrm{Aut}(\mathfrak{X})$ *and suppose that* G *is not compact. Then* G *acts transitively on* Ω *if and only if there exists* $x\in\mathfrak{X}$ *such that* $G\cap K_x$ *acts transitively on* Ω.

PROOF. Observe that, if G contains no translation, then by (8.1) it either is compact or fixes a point of the boundary. Therefore, under the hypothesis that G is noncompact and acts transitively on Ω, there exist a geodesic γ and an element $\tau\in G$ which is a translation along γ. Choose $x\in\gamma$. The subgroup $K_x\cap G$ is open in G and has countable index. This means that, if $\omega\in\Omega$, then, by the transitive action of G, $\Omega=G(\omega)=\bigcup_i h_i(G\cap K_x)(\omega)$, where h_i is a complete set of coset representatives. Since Ω is a complete metrizable space and $h_i(G\cap K_x)(\omega)$ is compact, it follows that $h_i(G\cap K_x)(\omega)$ has an interior point. Therefore the orbit $(G\cap K_x)(\omega)$ has an interior point, and hence it is open. But if the orbits of $G\cap K_x$ in Ω are open, they must be finitely many. Let ω' and ω'' be the two ends of γ. Then the orbits of ω' and ω'' under $G\cap K_x$ are open. This means that there exist $x',x''\in\gamma$, with $x'\in[x,\omega')$ and $x''\in[x,\omega'')$ such that $\{\omega\colon x'\in[x,\omega)\}\subseteq(K_x\cap G)(\omega')$ and $\{\omega\colon x''\in[x,\omega)\}\subseteq(K_x\cap G)(\omega'')$ (Fig. 14).

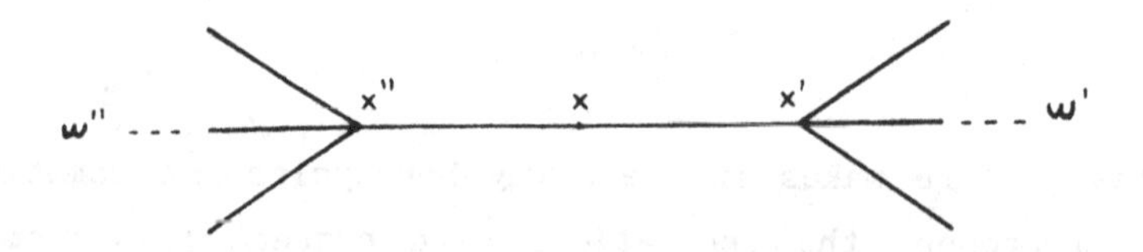

Fig. 14

But G contains a translation τ on the geodesic (ω',ω''). Suppose that τ moves x in the direction of ω''. Then there exists an index N such that, for $n\geq N$, $x''\in[x,\tau^n x]$. Therefore, if $\omega\neq\omega'$, we have $[x,\tau^n\omega)=[x,\tau^n x]\cup[\tau^n x,\tau^n\omega)$ and $\tau^n\omega\in(K_x\cap G)(\omega'')$. Let $k\in K_x\cap G$ be such that $k\tau^n(\omega)=\omega''$. Then $[x,k\tau^n\omega)=[x,k\tau^n x)\cup[k\tau^n x,\omega'')$. This implies that $k\tau^n x=\tau^n x$. Thus $\tau^{-n}k\tau^n x=x$ and $\tau^{-n}k\tau^n\in K_x\cap G$. But $\tau^{-n}k\tau^n(\omega)=\omega''$, and therefore $\omega\in(K_x\cap G)(\omega'')$. We have shown that $\Omega\setminus\{\omega'\}\subseteq(K_x\cap G)(\omega'')$. Similarly $\Omega\setminus\{\omega''\}\subseteq(K_x\cap G)(\omega')$. But, if $q\geq 2$, then $\Omega\setminus\{\omega'\}$ and $\Omega\setminus\{\omega''\}$ intersect, which implies that $K_x\cap G$ acts transitively on Ω. Obviously, if $K_x\cap G$ acts transitively on Ω, for some x, then so does G. ∎

The next result shows that a group which acts transitively on Ω and is not compact has at most two orbits on $\mathfrak{X}$. These orbits are nothing but the equivalence classes of the relation "$\mathbf{d}(x,y)$ is an even number". The fact that this is an equivalence relation follows from the observation that, if $x,y,z,\in\mathfrak{X}$ and t is the point of $[x,y]$ of minimal distance from z, then $\mathbf{d}(x,y)=\mathbf{d}(x,z)+\mathbf{d}(z,y)-2\mathbf{d}(z,t)$. If we fix $\mathbf{o}\in\mathfrak{X}$, these equivalence classes may be written as $\mathfrak{X}^+=\{x:\mathbf{d}(\mathbf{o},x)$ is even$\}$ and $\mathfrak{X}^-=\{x:\mathbf{d}(x,\mathbf{o})$ is odd$\}$. With this notation we prove the following.

(10.2) *PROPOSITION. Let* G *be a closed noncompact subgroup of* Aut($\mathfrak{X}$) *which acts transitively on* Ω. *Then either* G *is transitive on the vertices, or* G *has exactly the orbits* $\mathfrak{X}^-$ *and* $\mathfrak{X}^+$.

PROOF. By (10.1) there exists x such that $G\cap K_x$ acts transitively on Ω. Therefore $G\cap K_x$ acts transitively on the sets $\mathfrak{B}_n^x=\{y:\ \mathbf{d}(x,y)=n\}$, for every n. Let $y\in G(x)$, and let $n=\mathbf{d}(x,y)$. Then $\mathfrak{B}_n^x\subseteq G(x)$. Therefore G(x) is the union $G(x)=\bigcup_{n\in E}\mathfrak{B}_n^x$, where E is a set of nonnegative integers. Let k be the smallest integer other than zero occurring in E, and let $y\in\mathfrak{B}_k^x$. Then $y\in G(x)$. Furthermore K_y is a conjugate subgroup of $K_x\cap G$. Therefore $K_y\cap G$ acts transitively on Ω and on the sets $\mathfrak{B}_n^y=\{z:\ \mathbf{d}(z,y)=n\}$. This means that every element having distance k from y belongs to

$G(x)$. It follows that E must contain all multiples of k. We show now that k is at most 2. If $k=1$ then $G(x)=\mathfrak{X}$ and G acts transitively on $\mathfrak{X}$. Suppose now that $k>1$. Let $y\in\mathfrak{B}^x_k$, and let x' be the first element after x in the chain [x,y]. Let z be a vertex, not in the chain [x,y], having distance 2 from x and distance 1 from x' (Fig.15).

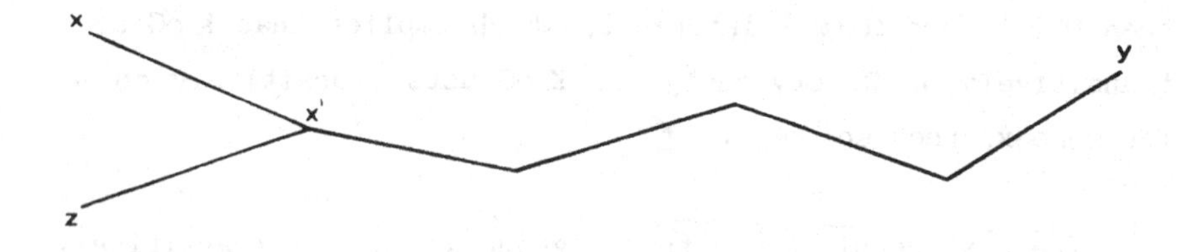

Fig.15

Then $\mathbf{d}(z,y)=k$ and therefore $z\in\mathfrak{B}^y_k\subseteq G(x)$. This implies that $\mathfrak{B}^x_2\subseteq G(x)$ and therefore $k=2$. Since $\bigcup_{j\geq 0}\mathfrak{B}^x_{kj}\subseteq G(x)$, we conclude that $G(x)=\{v\in\mathfrak{X}\colon \mathbf{d}(x,v) \text{ is even}\}$. By the proof of (10.1), it follows that there exists $t\in\mathfrak{X}$ with $\mathbf{d}(x,t)=1$ such that $G\cap K_t$ acts transitively on Ω. A similar argument proves that $G(t)=\{v\in\mathfrak{X}\colon \mathbf{d}(t,v) \text{ is even}\}$. Since $k>1$ and $\mathbf{d}(x,t)=1$ it follows that $G(x)\cap G(t)=\emptyset$, and the proposition follows. ∎

A notable subgroup of $\mathrm{Aut}(\mathfrak{X})$ which acts transitively on Ω, but not on $\mathfrak{X}$, is the group $\mathrm{Aut}(\mathfrak{X})^+$, generated by all rotations. The next result concerns this group.

(10.3) *PROPOSITION. The orbits of the group* $\mathrm{Aut}(\mathfrak{X})^+$ *generated by all rotations are exactly the equivalence classes* $\mathfrak{X}^+$ *and* $\mathfrak{X}^-$ *of the equivalence relation "*$\mathbf{d}(x,y)$ *is even". Furthermore, if* x *is any vertex,* $\mathrm{Aut}(\mathfrak{X})^+=K_x\Gamma^+$, *where* Γ *is any faithful transitive subgroup and* Γ^+ *is the subgroup of* Γ *which leaves invariant* $\mathfrak{X}^+$

and $\mathfrak{X}^-$. *Finally* $\mathrm{Aut}(\mathfrak{X})^+$ *is the only noncompact proper subgroup of* $\mathrm{Aut}(\mathfrak{X})$ *which contains* K_x.

PROOF. Observe that the $\mathfrak{X}^+$ and $\mathfrak{X}^-$ may be defined with reference to any vertex **o**. Clearly they are invariant under the group of rotations about **o**. Since **o** is arbitray they are invariant under any rotation. By (5.2), $\mathrm{Aut}(\mathfrak{X})^+$ is noncompact, and therefore by (10.2) its orbits are exactly $\mathfrak{X}^+$ and $\mathfrak{X}^-$. Let $\mathrm{Aut}(\mathfrak{X})=K_x\Gamma$, where Γ is a faithful transitive subgroup (6.3). Let $\Gamma^+=\{g\in\Gamma: \mathbf{d}(x,g(x))$ is even$\}$. Let $g\in\Gamma^+$, then there exists $h\in\mathrm{Aut}(\mathfrak{X})^+$ such that $g(x)=h(x)$. Thus $g^{-1}h\in K_x$, and $g\in\mathrm{Aut}(\mathfrak{X})^+$. Therefore $\Gamma^+\subseteq\mathrm{Aut}(\mathfrak{X})^+$. Since Γ acts transitively on $\mathfrak{X}$, so does Γ^+ on $\mathfrak{X}^+$. This means that a similar argument shows that $\mathrm{Aut}(\mathfrak{X})^+=K_x\Gamma^+$. Let now G be a noncompact proper subgroup of $\mathrm{Aut}(\mathfrak{X})$, containing K_x. Then, by (10.2), $G(x)=\mathfrak{X}$, or $G(x)=\mathfrak{X}^+$. In the first case, if $g\in\Gamma$, there exists $h\in G$, such that $g(x)=h(x)$, which implies $g^{-1}h\in K_x$ and $g\in G$. Thus G contains every faithful transitive subgroup and it must be all of $\mathrm{Aut}(\mathfrak{X})$. In the second case G contains Γ^+ and therefore $G=\mathrm{Aut}(\mathfrak{X})^+$ because $\mathrm{Aut}(\mathfrak{X})^+$ has index 2 in $\mathrm{Aut}(\mathfrak{X})$ (in fact $\mathrm{Aut}(\mathfrak{X})^+=\{g\in\mathrm{Aut}(\mathfrak{X}): \mathbf{d}(x,g(x))$ is even$\}$). ∎

We remark that the proposition above implies that $\mathrm{Aut}(\mathfrak{X})^+$ is generated by K_x and a nontrivial rotation not belonging to K_x. In particular, if $\{a,b\}$ is an edge, $\mathrm{Aut}(\mathfrak{X})^+$ is generated by the subgroups K_a and K_b. It can actually be shown that $\mathrm{Aut}(\mathfrak{X})^+$ is, in a natural way, the amalgamated product of K_a and K_b over their intersection $K_a\cap K_b$. The relationship between groups acting on a tree and amalgamated products is studied in **[Se]** where the reader will find the proof of the statement above.

A group G is said to act doubly transitively on $\mathfrak{X}$ if, for every two pairs (x,y), $(z,t)\in\mathfrak{X}\times\mathfrak{X}$ with $\mathbf{d}(z,t)=\mathbf{d}(x,y)$, there exists an element g of the group such that $g(x)=z$ and $g(y)=t$. A closed subgroup G of $\mathrm{Aut}(\mathfrak{X})$ is doubly transitive on $\mathfrak{X}$, if and only if G is transitive on $\mathfrak{X}$ and, for every x, $G\cap K_x$ acts transitively on the sets $\mathfrak{B}^x_n=\{y: \mathbf{d}(x,y)=n\}$. Equivalently, a closed subgroup G acts doubly transitively on $\mathfrak{X}$ if and only if it acts transitively on $\mathfrak{X}$ and Ω. There is an analogous notion of doubly transitive action on Ω. A closed subgroup G of Aut($\mathfrak{X}$) acts doubly transitively on Ω if and only if G acts transitively on Ω and $G \cap G_\omega = \{g \in G: g(\omega)=\omega\}$ acts transitively on $\Omega \setminus \{\omega\}$, for every $\omega \in \Omega$. We observe that $\mathrm{Aut}(\mathfrak{X})^+$ is doubly transitive on Ω because, for each ω, it contains the group B_ω which acts transitively on $\Omega \setminus \{\omega\}$ and because it contains a full group of rotations about a vertex, which is transitive on Ω. Nevertheless $\mathrm{Aut}(\mathfrak{X})^+$ is not transitive on $\mathfrak{X}$.

We conclude this chapter with a necessary condition for a group of automorphisms to contain faithful transitive subgroups of every isomorphism type. We observe first that a group G is transitive on $\mathfrak{E}$, the set of edges, if it is transitive on $\mathfrak{X}$ and, for every x, $K_x \cap G$ is transitive on $\mathfrak{B}_1^x = \{y: \mathbf{d}(x,y)=1\}$. With this in mind we can prove the following proposition.

(10.4) *PROPOSITION. Let* G *be a closed subgroup of* Aut($\mathfrak{X}$) *and suppose that*

(a) G *acts transitively on the set of edges* $\mathfrak{E}$,

(b) G *contains an inversion of order* 2.

Then, for all integers t, s *such that* $2t+s=q+1$, G *contains a faithful transitive subgroup isomorphic to the free product of* t *copies of* $\mathbb{Z}$ *and* s *copies of* $\mathbb{Z}_2$.

PROOF. Since G is transitive on the edges, condition (b) implies that there are inversions of order 2 on every edge. Therefore, by (6.3), it is sufficient to show that, for any three vertices x, x', x'', with $x' \neq x''$ and $\mathbf{d}(x,x')=\mathbf{d}(x,x'')=1$, there exists $g \in G$ such that $g(x')=x$ and $g(x)=x''$, in other words G contains a step-1 translation on a geodesic containing $[x',x'']$. By condition (a) there exists $\psi \in G$ such that $\psi(x)=x$ and $\psi(x')=x''$, and there exists $\phi \in G$ such that $\phi(x')=x$ and $\phi(x)=x'$. Thus $g=\psi\phi^{-1} \in G$, $g(x')=x$ and $g(x)=x''$. ∎

11. Notes and Remarks. Homogeneous trees and their automorphism groups come up naturally in many areas of mathematics. The interest in trees over the past 15 years was kindled by the lecture notes by J.P.Serre **[Se]** which were made available as mimeographed notes well before their actual publication. P.Cartier initiated the study of spherical functions on trees **[C3]** (see Chapter II, below).

Many of our definitions and simple geometrical ideas on trees are taken from the work of P.Cartier **[C1,C2]**. The contributions of J.Tits to the study of groups of automorphisms of trees are very important. In his paper **[Ti1]** he shows that the group $\mathrm{Aut}(\mathfrak{X})^{+}$ is simple. The simple classification of the automorphisms of a tree given in Section 3 and Lemma (5.3) are taken from **[Ti1]**. Theorem (6.3) is taken from **[Ch]** which in turn is based on **[F-T P3]**. An earlier result of the same type is contained in **[BP]**. The characterization of amenable groups given by Theorem (8.3) is taken from **[N4]**, where, as in these notes, it is deduced from Theorem (8.1) which is due to J.Tits **[Ti1]**. The fact that a solvable subgroup of $\mathrm{Aut}(\mathfrak{X})$ satisfies one of the conditions (*i*), (*ii*) and (*iii*) of Theorem (8.3) also follows from a result of **[Ti2]** concerning more general trees ($\mathbb{R}$-trees). A characterization of amenable subgroups of $\mathbf{PSL}(2,\mathbb{R})$ similar to that of Theorem (8.3) is given in **[N5]**.

Horocycles on trees were introduced by P.Cartier **[C2]**. The results of Section 10 are taken from **[N2]** and **[N1]**. The condition (b) of Proposition (10.4) is not necessarily satisfied by every group acting transitively on $\mathfrak{X}$ and Ω. The existence of counterexamples was kindly communicated to us by T. Steger. We describe here a counterexample under the hypothesis $q>2$. A similar more complicated example for the case $q=2$ was also found by T. Steger.

We consider the collection Σ of all subsets $\sigma\subseteq\mathfrak{X}$ consisting of exactly q vertices all adjacent to the same fixed vertex. For each $\sigma\subseteq\mathfrak{X}$ we choose once and for all a labelling, that is a coordinate mapping $\mathcal{T}_{\sigma}$: $\{1,2,\ldots\ldots,q\}\rightarrow\sigma$. Every automorphism of $\mathfrak{X}$ maps Σ into Σ and the map $\mathcal{T}^{-1}_{g(\sigma)}\circ g\circ\mathcal{T}_{\sigma}$ defines a permutation on q objects, i.e. an element of the symmetric group S(q). We say that $g\in\mathrm{Aut}(\mathfrak{X})$ is *$\mathcal{T}$-even* if the permutation $\mathcal{T}^{-1}_{g(\sigma)}\circ g\circ\mathcal{T}_{\sigma}$ is even for every $\sigma\in\Sigma$. In order to define an *$\mathcal{T}$-odd* automorphism we consider an edge $\{a,b\}$ and two elements $\sigma_a,\sigma_b\in\Sigma$ consisting respectively of all vertices adjacent to a except b, and all vertices adjacent to b except a. For g to be *$\mathcal{T}$-odd* we require that the permutations $\mathcal{T}^{-1}_{g(\sigma_a)}\circ g\circ\mathcal{T}_{\sigma_a}$ and $\mathcal{T}^{-1}_{g(\sigma_b)}\circ g\circ\mathcal{T}_{\sigma_b}$ have different parity for every edge $\{a,b\}\in\mathfrak{E}$. We now let $G=H^{+}\cup H^{-}$ where H^{+} and H^{-} are defined as follows. The set H^{-} consists of all *$\mathcal{T}$-odd* elements $g\in\mathrm{Aut}(\mathfrak{X})$ such that $\mathbf{d}(g(x),x)$ is odd for every $x\in\mathfrak{X}$ (the latter condition implies that g is either an inversion or an odd-step translation). The set H^{+} (which is a subgroup) consists of all *$\mathcal{T}$-even* elements g such that $\mathbf{d}(g(x),x)$ is even for every $x\in\mathfrak{X}$ (this means that g is a rotation or an even-step translation). It is easy to verify that G is a closed subgroup without inversions of order 2. To show that G is transitive on Ω it suffices to prove that, given $x,y\in\mathfrak{X}$ at equal distance from a vertex $\mathbf{o}\in\mathfrak{X}$, there exists an element of H^{+} which maps x into y leaving $\mathbf{o}$ fixed. This element may be constructed step by step so that the condition of belonging to H^{+} is respected; it is in this step-by-step construction that the condition $q+1>3$ is used. A similar construction allows us to find a translation of odd step in H^{-} and this is sufficient (by Proposition (10.2)) to prove that G acts transitively on $\mathfrak{X}$.

The counterexamples and Theorem (6.3) imply that if q+1 is odd then there exist closed subgroups of $\mathrm{Aut}(\mathfrak{X})$ acting transitively on $\mathfrak{X}$ and Ω which do not contain faithful transitive subgroups. If q+1 is even, condition (a) of (10.4) is enough for the existence of a faithful transitive subgroup isomorphic to a free group with (q+1)/2 generators. On the other hand, F. Bouaziz-Kellil **[BK2,** Ex. f, p.19] gives for every q an example of a discrete subgroup acting transitively on $\mathfrak{X}$ which fails to contain a faithful transitive subgroup.

CHAPTER II

1. Eigenfunctions of the Laplace operator. Let f be a function defined on $\mathfrak{X}$, the set of vertices of a homogeneous tree. We let Lf(x) denote the *average of the values of* f *on the nearest neighbors of* x. In other words,

$$Lf(x) = (q+1)^{-1} \sum_{\{x,y\}\in\mathfrak{E}} f(y) .$$

The operator L which is defined on the vector space of all complex-valued functions on $\mathfrak{X}$ is called the Laplace operator on $\mathfrak{X}$. (Sometimes, perhaps more appropriately, the name Laplace operator is given to the operator (Lf−f).)

In this section we will characterize the eigenfunctions of the Laplace operator, that is, the functions f which satisfy $Lf=\mu f$, for some eigenvalue $\mu\in\mathbb{C}$.

First of all we define certain *elementary* eigenfunctions associated to points of the boundary.

Recall that if [x,y] is any (oriented) chain the sets $\Omega(x,y)=\{\omega\colon$ [x,y] is a subchain of $[x,\omega)\}$ are open in Ω, and indeed form a basis. If $x\in\mathfrak{X}$, the sets $\Omega(x,t)$ with $\mathbf{d}(x,t)=n$ form a partition of Ω into $(q+1)q^{n-1}$ disjoint open and closed sets. Therefore there exists a unique Borel probability measure ν_x on Ω such that

$$\nu_x(\Omega(x,t)) = \frac{q}{q+1}\, q^{-\mathbf{d}(x,t)} .$$

The measures ν_x are all absolutely continuous with respect to each other. This is most easily seen by computing the Radon-Nikodym derivative $d\nu_y/d\nu_x(\omega)$, which will be shown to be *an everywhere positive function of* ω *assuming only finitely many values.*

Let $\omega\in\Omega$ and $x,y\in\mathfrak{X}$. Let $\{x,s_1,s_2,\dots\}=[x,\omega)$ be the infinite chain from x to ω. The sets $\Omega(x,s_n)$ are a basic system of neighborhoods of ω, and $\nu_x(\Omega(x,s_n))=(q+1)^{-1}q^{-n+1}$.

Consider the sequence of horocycles of ω, $\{H_n: n\in\mathbb{Z}\}$, indexed so that $x\in H_0$ (Fig.1), and $s_n\in H_n$, for $n>0$. Let $y\in\mathfrak{X}$ and suppose $y\in H_k$, $k\in\mathbb{Z}$. Choose $n>0$ and $n>k$. Then $\nu_y(\Omega(x,s_n))= \nu_x(\Omega(x,s_n))q^k$. Letting n go to infinity, we obtain $d\nu_y/d\nu_x(\omega)= q^k$.

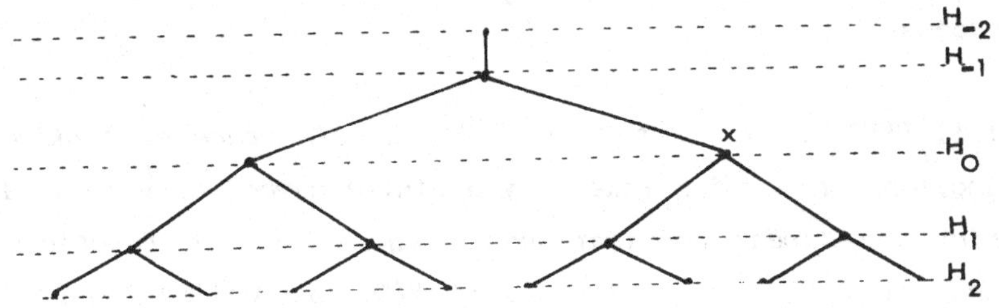

Fig.1

To give a more concise description of the Radon-Nikodym derivative, we introduce the function $\delta(x,y,\omega)$ which is defined as the distance of the ω-horocycle H(x) containing x and the ω-horocycle H(y) containing y, taken with the positive sign if H(y) is closer to ω than H(x), and with the negative sign otherwise. With this notation $d\nu_y/d\nu_x(\omega)=q^{\delta(x,y,\omega)}$.

Observe that, for fixed x and y, the function $\delta(x,y,\omega)$ takes values in $\mathbb{Z}$, and $-\mathbf{d}(x,y)\le\delta(x,y,\omega)\le\mathbf{d}(x,y)$. This means that $d\nu_y/d\nu_x(\omega)$ can assume only finitely many values. We shall use the notation

$$P(x,y,\omega) = \frac{d\nu_y}{d\nu_x}(\omega).$$

Let now $\mathbf{o}$ be a fixed vertex, ω an element of Ω, and $z\in\mathbb{C}$. Define $f_z(x)=P^z(\mathbf{o},x,\omega)$. We shall prove that

$$Lf_z(x) = \mu(z)\, f_z(x),$$

where $\mu(z)=\dfrac{q^z+q^{1-z}}{q+1}$.

If t is a nearest neighbor of x, then $\delta(\mathbf{o},t,\omega)=\delta(\mathbf{o},x,\omega)\pm1$.

As t varies among the q+1 nearest neighbohrs of x, the sign + occurs once and the sign - occurs q times. Therefore

$$(q+1)^{-1} \sum_{\{x,t\}\in\mathfrak{E}} f_z(t) = (q+1)^{-1}(f_z(x)q^z + f_z(x)\, q^{1-z}) = \mu(z)f_z.$$

This shows that f_z is an eigenfunction. We will show that every eigenfunction with eigenvalue $\mu(z)$ can be obtained from the elementary eigenfunctions f_z by an appropriate limiting process.

(1.1) *DEFINITION.* *Let $\mathcal{K}(\Omega)$ be the linear space of continuous functions on Ω which take only a finite number of values, i.e. which are linear combinations of characteristic functions of the sets $\Omega(\mathbf{o},x)$. The elements of $\mathcal{K}(\Omega)$ are called cylindrical functions. Let $\mathcal{K}'(\Omega)$ be the space of finitely additive complex-valued set functions defined on the algebra of subsets of Ω generated by the open and compact sets $\Omega(\mathbf{o},x)$. The elements of $\mathcal{K}'(\Omega)$ are called finitely additive measures.*

It is not difficult to show that $\mathcal{K}'(\Omega)$ is the dual space of $\mathcal{K}(\Omega)$. Indeed the sets of constancy of an element of $\mathcal{K}(\Omega)$ belong to the algebra of subsets of Ω which is the domain of the elements of $\mathcal{K}'$.

Observe in particular that $P(\mathbf{o},x,\omega)\in\mathcal{K}$, if $\mathbf{o}\in\mathfrak{X}$ is fixed. It makes sense therefore, for $z\in\mathbb{C}$, to define, for $m\in\mathcal{K}'$,

$$\mathcal{P}_z m(x) = \int_\Omega P^z(\mathbf{o},x,\omega)\, dm(\omega).$$

$\mathcal{P}_z m(x)$ is a $\mu(z)$-eigenfunction of L. This fact is easily proved by observing that, if x varies in a finite set,

$$\mathcal{P}_z m(x) = \sum_i m(A_i)\, P^z(\mathbf{o},x,\omega_i),$$

where the $A_i\subseteq\Omega$ are sets on which the functions $P^z(\mathbf{o},x,\omega)$ are constant and $\omega_i\in A_i$. The assertion then follows from the fact that, for fixed ω, $P^z(\mathbf{o},x,\omega)$ as a function of x is a

$\mu(z)$-eigenfunction. The following theorem says that every μ-eigenfunction may be expressed as $\mathcal{P}_z m(x)$, for some $m\in\mathcal{K}'$ and some $z\in\mathbb{C}$, such that $\mu(z)=\mu$. It is important to observe that the map $z\to\mu(z)$ is surjective.

(1.2) *THEOREM. Let* f *be a function defined on* $\mathfrak{X}$ *and satisfying* $Lf=\mu f$, *for some* $\mu\in\mathbb{C}$. *Let* $o\in\mathfrak{X}$, *and let* z *be a complex number such that* $\mu=\mu(z)=(q^z+q^{1-z})/(q+1)$, *and* $z\neq k i\pi/\ln q$, *for* $k\in\mathbb{Z}$. *Then there exists* $m\in\mathcal{K}'$ *such that*

$$f(x) = \mathcal{P}_z m(x) = \int_\Omega P^z(o,x,\omega)\, dm(\omega).$$

PROOF. For simpler notation, we shall write $P(x,\omega)$ in the place of $P(o,x,\omega)$. Let $\mathfrak{I}$ be a finite subtree of $\mathfrak{X}$, containing o. We say that a vertex of $\mathfrak{I}$ is an *interior point* if each of its $q+1$ nearest neighbors lies in $\mathfrak{I}$. A vertex of $\mathfrak{I}$ which is not an interior point is said to belong to the boundary $\partial\mathfrak{I}$ of $\mathfrak{I}$, or to be a *boundary point*. We call a complex-valued function on $\mathfrak{I}$ a μ-eigenfunction, if $Lf(x)=\mu f(x)$, *for every interior point* $x\in\mathfrak{I}$. Observe that, if $\mathfrak{I}$ has no interior point, then every function on $\mathfrak{I}$ is a μ-eigenfunction. Now every boundary point of $\mathfrak{I}$ is the last element of the chain $[o,\omega)\cap\mathfrak{I}$ for some $\omega\in\Omega$. Conversely, every $\omega\in\Omega$ identifies uniquely the last element y_ω of the chain $[o,\omega)\cap\mathfrak{I}$. The restriction to $\mathfrak{I}$ of the function $\delta(o,x,\omega)$ may be written as $\mathbf{d}(o,y_\omega)-\mathbf{d}(x, y_\omega)$. It follows that the restriction to $\mathfrak{I}$ of the function $P(x,\omega)$ may be written as

$$P(x,\omega)\Big|_{\mathfrak{I}} = q^{d(o,y_\omega)-d(x,y_\omega)} .$$

Therefore the functions $P^z(x,y)=q^{z(d(o,y)-d(x,y))}$, with $y\in\partial\mathfrak{I}$, are exactly the restrictions to $\mathfrak{I}$ of the functions $P^z(x,\omega)$. These functions are μ-eigenfunctions on $\mathfrak{I}$, because, as shown before the statement of the theorem, $P^z(x,\omega)$ is a μ-eigenfunction on $\mathfrak{X}$, for every $\omega\in\Omega$. We shall prove that the set $B=\{P^z(x,y)\colon y\in\partial\mathfrak{I}\}$ is a basis for the vector space of the μ-eigenfunctions on $\mathfrak{I}$. We prove first that it is linearly independent. We use induction on the number of vertices in $\mathfrak{I}$.

Assume that **B** is linearly independent, and let $\mathfrak{T}'$ be the tree obtained from $\mathfrak{T}$ by adding one point. If the new point is added in such a way that some boundary point of $\mathfrak{T}$ becomes an interior point of $\mathfrak{T}'$, then the number of elements of $\partial\mathfrak{T}'$ remains the same. Furthermore, if y is the element of $\partial\mathfrak{T}$ which becomes an interior point and y' is the added boundary point then y' is the last element of the chain $[\mathbf{o},\omega)\cap\mathfrak{T}'$ if and only if y is the last element of the chain $[\mathbf{o},\omega)\cap\mathfrak{T}$. Therefore, for every $x\in\mathfrak{T}$, the element $P(x,y)$ of **B** has the same value as the element $P(x,y')$ of **B'**. This means that the induction hypothesis implies that the restrictions to $\mathfrak{T}$ of the elements of **B'** are linearly independent. In this case the elements of **B'** are *a fortiori* independent of $\mathfrak{T}'$. If each point of $\partial\mathfrak{T}$ is also a point of $\partial\mathfrak{T}'$, let $\partial\mathfrak{T}'=\{t,y_1,\dots,y_n\}$, where $\partial\mathfrak{T}=\{y_1,\dots y_n\}$, and t is adjacent to y_1. If, for $x\in\mathfrak{T}'$,

$$c_0P^z(x,t) + \sum_{i=1}^{n} c_iP^z(x,y_i) = 0,$$

then the same holds for $x\in\mathfrak{T}$, where $P^z(x,y_1)=P^z(x,t)$. It follows that, for $x\in\mathfrak{T}$,

$$(c_0+c_1)P^z(x,y_1)+ \sum_{i=2}^{n} c_iP^z(x,y_i) = 0.$$

This implies, by the induction hypothesis, $c_0+c_1=0$, and $c_2= c_3=\dots\dots=c_n=0$. Therefore $c_0P^z(x,t)+c_1P^z(x,y_1)=0$, for $x\in\mathfrak{T}'$. In particular,

$$c_0P^z(t,t) + c_1P^z(t,y_1) = 0 .$$

The chain $[\mathbf{o},y_1]$ is contained in $\mathfrak{T}$, and therefore does not include t. But y_1 and t are nearest neighbors, therefore the chain $[\mathbf{o},t]$ contains y_1. Therefore $P^z(t,t)=q^{zd(\mathbf{o},t)}$ and $P^z(t,y_1) = q^{zd(\mathbf{o},y_1)-zd(t,y_1)} = q^{z(d(\mathbf{o},t)-2)}$. It follows that $c_0+c_1q^{-2z}=0$, which together with $c_0+c_1=0$ is possible only with $c_0=c_1=0$, unless $q^{-2z}=q^{2z}=1$. The case $q^{2z}=1$, that is $z=ik\pi/\ln q$, with $k\in\mathbb{Z}$, has been excluded by hypothesis. We conclude that $c_0=c_1=0$, and the set **B** is linearly independent. We will prove now that the space of μ-eigenfunctions on $\mathfrak{T}$ has dimension n, where n is the number of elements of $\partial\mathfrak{T}=\{y_1,\ldots,y_n\}$. Again we use induction on the size of $\mathfrak{T}$. If $\mathfrak{T}'$ is obtained by adding one point, say t, to $\mathfrak{T}$, then t must be adjacent to some point $y_1\in\partial\mathfrak{T}$. It follows that either y_1 remains a boundary point in $\mathfrak{T}'$, or it becomes an interior point. In the first case a μ-eigenfunction on $\mathfrak{T}$ may be extended to a μ-eigenfunction on $\partial\mathfrak{T}'$, by assigning an arbitrary value at t, because the condition of being a μ-eigenfunction applies only to interior points and t is not adjacent to an interior point. In the second case the number of boundary points remains the same and a μ-eigenfunction f on $\mathfrak{T}$ extends uniquely to the point t. In fact the unique extension is

$$f(t) = (q+1)\mu f(y_1) - \sum_{\substack{\{x,y_1\}\in\mathfrak{E} \\ x\neq t}} f(x) .$$

This shows that the dimension of the space of μ-eigenfunctions on $\mathfrak{T}$ is the same as the number of points of $\partial\mathfrak{T}$. Thus, if $\mathfrak{T}$ is any subtree containing **o**, the subset **B** is a basis for the space of μ-eigenfunctions of L on $\mathfrak{X}$. Let $\mathfrak{X}_n=\{x\in\mathfrak{X}: \mathbf{d}(\mathbf{o},x)\leq n\}$. Then $\mathfrak{X}_n$ is a subtree with boundary $\partial\mathfrak{X}_n=\{x\in\mathfrak{X}: \mathbf{d}(\mathbf{o},x)=n\}$. By what we have just proved $f|_{\mathfrak{X}_n}$ can be expressed uniquely as

$$f(x) = \sum_{d(y,o)=n} m_y\, P^z(x,y),$$

for some complex numbers m_y. The numbers m_y are uniquely determined and we may define $m(\Omega(\mathbf{o},y))=m_y$. This definition determines a finitely additive measure, because

$$m_y = \sum m_t$$

where the sum is taken over all t such that $\mathbf{d}(\mathbf{o},t)>\mathbf{d}(\mathbf{o},y)$, and $\{y,t\}\in\mathfrak{E}$. By definition, for each x

$$f(x) = \int_\Omega P^z(x,\omega)dm(\omega) .$$

The uniqueness of the numbers $\{m_y\}$ implies that the measure m is unique.∎

The relationship between the eigenfunction f and the finitely additive measure m for a given eigenvalue $\mu(z)$ is given by the following proposition. We write for simplicity $P(\mathbf{o},x,\omega)=P(x,\omega)$, and $\Omega(\mathbf{o},x)=\Omega(x)$.

(1.3) *PROPOSITION. Let* f *be a* μ*-eigenfucntion with* $\mu=\mu(z)=(q^z+q^{1-z})(q+1)^{-1}$, *and let* $m\in\mathcal{K}'(\Omega)$ *be the finitely additive measure such that*

$$f(x) = \int_\Omega P^z(x,\omega)\ dm(\omega).$$

Then $m(\Omega)=f(\mathbf{o})$. *Moreover, if* x' *is the vertex at distance* 1 *from* x *in the chain* $[\mathbf{o},x]$, *then*

$$m(\Omega(x)) = (q^z-q^{-z})^{-1}q^{-zd(\mathbf{o},x')}(f(x) - q^zf(x')).$$

PROOF. The fact that $m(\Omega)=f(\mathbf{o})$ follows directly from the fact that $P(\mathbf{o},x)=1$. Observe that $P^z(x,\omega)=q^{-z}P^z(x',\omega)$ if $\omega\notin\Omega(x)$, and $P^z(x,\omega)=q^zP^z(x',\omega)$ if $\omega\in\Omega(x)$. Therefore

$$f(x)= \int_\Omega P^z(x,\omega)dm= \int_{\Omega\setminus\Omega(x)} P^z(x,\omega)dm + \int_{\Omega(x)} P^z(x,\omega)dm$$

$$= q^{-z}\int_{\Omega\setminus\Omega(x)} P^z(x',\omega)dm + q^z\int_{\Omega(x)} P^z(x',\omega)dm.$$

On the other hand $P^z(x',\omega)$ is equal to $q^{zd(x',\mathbf{o})}$ for every $\omega\in\Omega(x)$. Therefore,

$$\int_{\Omega(x)} P^z(x',\omega)dm = m(\Omega(x))q^{zd(x',\mathbf{o})},\ \text{while}$$

$$\int_{\Omega\setminus\Omega(x)} P^z(x',\omega)dm= \int_\Omega P^z(x',\omega)dm - \int_{\Omega(x)} P^z(x',\omega)dm$$

$$= f(x')-m(\Omega(x))q^{zd(x',\mathbf{o})}.$$

Therefore,

$$f(x)-q^{-z}(f(x')=(q^{z}-q^{-z})q^{zd(x',o)}) + q^{-zd(x',o)}m(\Omega(x)),$$

and $m(\Omega(x)=(q^{z}-q^{-z})q^{-zd(x',o)}(f(x)-q^{-z}f(x'))$.█

2. Spherical functions. As before let $o\in\mathfrak{X}$ be a fixed vertex. We shall call a function defined on $\mathfrak{X}$ *radial* (with respect to the vertex **o**) if it depends only on the distance from the vertex **o**. That is, f is radial if $d(o,x)=d(o,y)$ implies $f(x)=f(y)$. We study in this section the *radial eigenfunctions* of L.

As in the proof of (1.2), we write again for simplicity of notation $P(o,x,\omega)=P(x,\omega)$. Then by (1.2) a μ-eigenfuntion of L may be written as

$$f(x) = \int_{\Omega} P^{z}(x,\omega)\ dm\ ,$$

where m is a finitely additive measure and $\mu=\mu(z)$. We should therefore identify the finitely additive measures which yield *radial* eigenfunctions.

(2.1) *THEOREM. A μ-eigenfunction of* L *is radial if and only if it is a constant multiple of the function*

$$\phi_{z}(x) = \int_{\Omega} P^{z}(x,\omega)\ d\nu(\omega),$$

where $\mu=\mu(z)$ and $\nu=\nu_{o}$ is the positive measure on Ω defined by

$$\nu(\Omega)=1,\ \ \nu(\Omega(o,x))=q^{-d(o,x)}\frac{q}{q+1}.$$

PROOF. We observe first of all that the measure ν defined above is (up to multiplication by a constant) the only finitely additive measure which is invariant under $K=K_{o}$. That is $\nu(kE)=\nu(E)$, if E is a basic open set and $k\in K$. Next we observe that, if m is a finitely additive measure, the measure defined on the basic open sets by

$$m^{K}(E) = \int_{K} m(kE)\ dk\ ,$$

where dk is the normalized *Haar* measure on K, is K-invariant. Therefore $m^K=m(\Omega)\nu$. Recalling now the definition of $P(x,\omega)$, and $\delta(o,x,\omega)$, given in Section 1, we observe that $\delta(o,kx,k\omega)=\delta(o,x,\omega)$, provided that $ko=o$; hence $\delta(o,kx,\omega)=\delta(o,x,k^{-1}\omega)$, if $k\in K$, which implies $P(kx,\omega)=P(x,k^{-1}\omega)$. Thus

$$\phi_z(kx) = \int_\Omega P^z(x,k^{-1}\omega)\, d\nu = \int_\Omega P^z(x,\omega)\, d\nu = \phi_z(x),$$

because ν is K-invariant. On the other hand if f is a μ-eigenfunction which is K-invariant, then by (1.2)

$$f(x) = \int_\Omega P^z(x,\omega)\, dm(\omega) ,$$

and

$$f(x)= \int_K f(kx)dk= \int_\Omega\int_K P^z(x,k^{-1}\omega)dm(\omega)dk= \int_\Omega P^z(x,\omega)dm^K(\omega)$$

$$= m(\Omega)\phi_z(x). \blacksquare$$

(2.2) *DEFINITION. Let ϕ be a radial eigenfunction of* L *satisfying* $\phi(o)=1$. *Then ϕ is called a spherical function.*

With this terminology we have now that, if ϕ is a spherical function with eigenvalue μ, then $\phi=\phi_z$, for some z such that $\mu(z)=\mu$. We will now compute the values of ϕ_z explicitly. To simplify notation we write $|x|=d(o,x)$. A radial function is then a function of $|x|$ alone. Since $|x|$ takes only integral values, we may think of a spherical function as a function on the nonnegative integers. We therefore write $\phi_z(n)=\phi_z(x)$, when $|x|=n$. Observe that the value of ϕ_z on the elements having distance one from o is exactly $\mu(z)$. By definition $\phi_z(o)=1$. Therefore the following lemma allows us to compute all the values of ϕ_z by induction.

(2.3) LEMMA. *If ϕ_z is a spherical function, then, for* $n\geq1$,

$$\phi_z(n+1) = \frac{q+1}{q}\, \phi_z(1)\phi_z(n) - \frac{1}{q}\, \phi_z(n-1).$$

PROOF. If $|x|=n$, the equality $L\phi_z(x)=\mu(z)\phi_z(x)$ implies

$$\frac{1}{q+1}\,\phi_z(n-1) + \frac{q}{q+1}\,\phi_z(n+1) = \mu(z)\phi_z(n). \blacksquare$$

(2.4) *PROPOSITION. Let* $z\in\mathbb{C}$*, and let* $h_z(x)=q^{-z|x|}$*, then*

(i) if $q^{2z-1}\neq 1$*, then, for every* $x\in\mathfrak{X}$*,*

$$\phi_z(x) = c(z)\,h_z(x) + c(1-z)\,h_{1-z}(x),$$

where

$$c(z) = \frac{1}{q+1}\,(q^{1-z}-q^{z-1})(q^{-z}-q^{z-1})^{-1},$$

(ii) if $q^{2z-1}=1$*, then, for every* $x\in\mathfrak{X}$*,*

$$\phi_z(x) = (1+\frac{q-1}{q+1}|x|)\;h_z(x).$$

PROOF. Consider the system (in the unknowns c and c') $c+c'=1$, $cq^{-z}+c'q^{z-1}=\mu(z)$. If $q^{2z-1}\neq 1$, then the system is nonsingular. Let $c(z)$ and $c'(z)$ be the solutions of the system. Since $\mu(z)=\mu(1-z)$, we also have $c'(z)=c(1-z)$. The expression of the solution $c(z)$ is given in the statement. The function

$$\phi(x) = c(z)\,h_z(x) + c(1-z)\,h_{1-z}(x)$$

is radial, and satisfies $\phi(\mathbf{o})=1$, $\phi(x)=\mu(z)$ for $|x|=1$. In particular $L\phi(\mathbf{o})=\mu(z)=\mu(z)\phi(\mathbf{o})$. In order to show that $\phi(x)=\phi_z(x)$, for every x, it suffices to show that $L\phi(x)=\mu(z)\phi(x)$ for $x\neq\mathbf{o}$. Observe that if $x\neq\mathbf{o}$, and y varies among the nearest neighbors of x, h_z assumes q times the value $q^{-z}h_z(x)$, and once the value $q^z h_z(x)$. Therefore $Lh_z(x)=\frac{q^{1-z}+q^z}{q+1}h_z(x)=\mu(z)h_z(x)$. Since ϕ is a linear combination of h_z and h_{1-z}, and $\mu(z)=\mu(1-z)$, it follows that $L\phi(x)=\mu(z)\phi(x)$. We conclude that $\phi=\phi_z$. Assume now that $q^{2z-1}=1$, and let $k_z(x)=|x|h_z(x)$. Suppose that $x\neq\mathbf{o}$; then

$$\begin{aligned} Lk_z(x) &= (q+1)^{-1}(q^{1-z}(|x|+1)+q^z(|x|-1)h_z(x) \\ &= \mu(z)k_z + (1+\frac{q-1}{q+1}|x|)h_z(x). \end{aligned}$$

Then ϕ is a linear combination of h_z and k_z. Therefore

$L\phi(x)=\mu(z)\phi(x)$ for $x\neq\mathbf{o}$. But $L\phi(\mathbf{o})$ is the value of ϕ on the elements of distance 1 from $\mathbf{o}$, and therefore $L\phi(\mathbf{o})=q^{-z}(1+\frac{q-1}{q+1})=\mu(z)=\mu(z)\phi(\mathbf{o})$. This shows that $L\phi=\mu(z)\phi$, and, since ϕ is radial, $\phi=\phi_z$. ∎

3. Intertwining operators. Let $z\in\mathbb{C}$ be such that $q^{2z}\neq\pm 1$. Let $\mathbf{o}$ be a fixed vertex and, as before, write $P(\mathbf{o},x,\omega)=P(x,\omega)$ and $\Omega(\mathbf{o},x)=\Omega(x)$. Recall that ν is the unique probability measure on Ω, which is invariant under $K_{\mathbf{o}}=K$. Let $\mathcal{K}_n(\Omega)$ be the linear subspace of $\mathcal{K}(\Omega)$ generated by $\{\chi_{\Omega(x)}: \mathbf{d}(x,\mathbf{o})=n\}$. The *Poisson transform* $\mathcal{P}_z$ is defined on the space $\mathcal{K}'(\Omega)$ of finitely additive measures as

$$\mathcal{P}_z m(x) = \int_\Omega P^z(x,\omega)\, dm(\omega).$$

Observe that both $\mathcal{P}_z$ and $\mathcal{P}_{1-z}$ map the space $\mathcal{K}'$ onto the space of $\mu(z)$-eigenfunctions where $\mu(z)=(q^z+q^{1-z})(q+1)^{-1}$. This follows from (1.2) and our hypothesis on $z\in\mathbb{C}$.

(3.1) *DEFINITION. For* $z\in\mathbb{C}$ *satisfying the conditions above, and* $\xi\in\mathcal{K}_n$, *define* $I_z\xi\in\mathcal{K}_n$ *to be such that*

$$\mathcal{P}_z(I_z\xi d\nu)(x)= \int_\Omega P^z(x,\omega)I_z\xi(\omega)d\nu= \int_\Omega P^{1-z}(x,\omega)\xi(\omega)d\nu$$
$$= \mathcal{P}_{1-z}(\xi d\nu)(x).$$

To show that the definition makes sense we must prove that $I_z\xi\in\mathcal{K}_n(\Omega)$. But if $\xi\in\mathcal{K}_n$, then $\xi=\sum f(y)P^z(y,\omega)$, with f supported in $\mathfrak{X}_n=\{x: \mathbf{d}(\mathbf{o},x)\leq n\}$. Therefore

$$\int_\Omega P^{1-z}(x,\omega)\xi(\omega)d\nu= \sum f(y)\int_\Omega P^{1-z}(x,\omega)P^z(y,\omega)d\nu$$
$$= \int_\Omega P^z(x,\omega)\left[\sum f(y)P^{1-z}(y,\omega)\right]d\nu.$$

We have used the fact that the measure $P^{1-z}(x,\omega)P^z(y,\omega)d\nu$ has the same distribution as the corresponding measure where the roles of x and y are exchanged. Indeed the first measure may be written as $(d\nu_y/d\nu_x)^z d\nu_x$, which changing the role of x and y becomes $(d\nu_x/d\nu_y)^z d\nu_y$. Thus we can write $I_z\xi(\omega)=\sum f(y)P^{1-z}(y,\omega)$. In other words I_z may also be defined as the linear extension of the map $P^z(x,\omega) \to P^{1-z}(x,\omega)$. Observe that $I_z I_{1-z}$ is the identity operator. The operators I_z and I_{1-z} are called the *intertwining operators* relative to the eigenvalue $\mu(z)$.

We shall now find a common set of eigenvectors for the intertwining operators. Let $x\neq\mathbf{o}$ and let x' be the vertex of the chain $[\mathbf{o},x]$ which has distance 1 from x. Define $\xi_{\mathbf{o}}=1$ and, for $x\neq\mathbf{o}$, define

$$\xi_x=(q/q+1)(q^{-d(\mathbf{o},x)}\chi_{\Omega(x)} - q^{d(\mathbf{o},x')}\chi_{\Omega(x')}) .$$

(3.2) *LEMMA. The functions* ξ_x $x\in\mathfrak{X}$, *span* $\mathcal{K}(\Omega)$ *and, moreover,* $I_z\xi_{\mathbf{o}}=\xi_{\mathbf{o}}$, *and for* $x\neq\mathbf{o}$

$$I_z\xi_x=(q^{1-z}-q^{z-1})(q^z-q^{-z})^{-1}q^{-(2z-1)d(x',\mathbf{o})}\xi_x.$$

In particular, if $z=1/2+it$, I_z *extends to a unitary operator on* $L^2(\Omega,\nu)$, *and* $I_{1/2}$ *is the identity operator.*

PROOF. It is easy to see by induction that the set ξ_x , $\mathbf{d}(\mathbf{o},x)\le n$, spans $\mathcal{K}_n(\Omega)$. It is also obvious that $I_z\xi_{\mathbf{o}}=\xi_{\mathbf{o}}$. Let $\mathbf{d}(x,\mathbf{o})=n+1$, with $n\ge0$. Observe that $P^z(x,\omega)= \eta_1(\omega)+c_1\chi_{\Omega(x)}$ where $\eta_1\in\mathcal{K}_n$ and $c_1\neq0$. We can write therefore $\xi_x=\eta_2(\omega)+c_2P^z(x,\omega)$, with $\eta_2\in\mathcal{K}_n$. Therefore ξ_x may be written as $\xi_x(\omega)=\sum f(y)P^z(y,\omega)$, with f supported on $\{y:\ \mathbf{d}(y,\mathbf{o})\le n\}\cup\{x\}$. Since $I_z\xi_x(\omega)=\sum f(y)P^{1-z}(y,\omega)$, it follows that $I_z\xi_x= \eta_3(\omega)+c_3\xi_x$ where $\eta_3\in\mathcal{K}_n$. But if $\mathbf{d}(y,\mathbf{o})=n$ then $\int_{\Omega(y)}\xi_x d\nu=0$. Therefore $\mathcal{P}_z(\xi_x d\nu)$ and $\mathcal{P}_{1-z}(\xi_x d\nu)$ are identically zero on $\mathfrak{X}_n=\{y:\ \mathbf{d}(\mathbf{o},y)\le n\}$. So, for any $y\in\mathfrak{X}_n$,

$$\mathcal{P}_z(\eta_3 d\nu)(y)= \mathcal{P}_z(I_z\xi_x d\nu)(y)-c_3\mathcal{P}_z(\xi_x d\nu)(y)$$

$$= \mathcal{P}_{1-z}(\xi_x d\nu)-c_3\mathcal{P}_z(\xi_x d\nu)(y)=0.$$

This implies that $\int_{\Omega(y)}\eta_3(\omega)d\nu=0$ for every $y\in\mathfrak{X}_n$. But $\eta_3\in\mathcal{K}_n$, and therefore $\eta_3=0$. We have proved that $I_z\xi_x=c_3\xi_x$. It remains to compute c_3. Let $\psi(y)=\mathcal{P}_{1-z}(\xi_x d\nu)(y)=\mathcal{P}_z(I_z\xi_x d\nu)(y)$. Using

(3.1) and the fact that $\psi(y)=0$ for $d(y,o)\le n$, we compute

$$\int_\Omega I_z\xi_x d\nu=(q^z-q^{-z})^{-1}q^{-zd(o,x')}(\psi(x)-q^{-z}\psi(x'))$$
$$=(q^z-q^{-z})^{-1}q^{-zd(o,x')}\psi(x),\quad\text{and}$$
$$\int_\Omega \xi_x d\nu=(q^{1-z}-q^{z-1})^{-1}q^{-(1-z)d(o,x')}\psi(x).$$

It follows that $c_3=(q^{1-z}-q^{z-1})(q^z-q^{-z})^{-1}q^{-(2z-1)d(x',o)}$, and the proof of the first assertion is complete. The last assertion now follows when we observe that if $z=1/2+it$ the coefficients multiplying ξ_x are of modulus 1 and are 1 if $z=1/2$. ∎

4. The Gelfand pair (G,K). Let G be a closed subgroup of $\mathrm{Aut}(\mathfrak{X})$ and suppose that G acts transitively on $\mathfrak{X}$. If $o\in\mathfrak{X}$ is a fixed vertex, then the orbit Go is all of $\mathfrak{X}$. Therefore $\mathfrak{X}$ may be identified through the map $g\to go$ with the quotient G/K, where $K=G\cap K_o=\{g\in G:\ go=o\}$. This means that every function on $\mathfrak{X}$ may be lifted to a function on G, by defining $\tilde{f}(g)=f(go)$. The function $\tilde{f}$ has the property that $\tilde{f}(gk)=\tilde{f}(g)$, for every $k\in K$, and conversely a K-right-invariant function on G may be identified with a function on $\mathfrak{X}$.

If f is a measurable function on G then

$$f^K(g) = \int_K f(gk)\, dk$$

is a K-right-invariant function on G, and the map $f\to f^K$ preserves continuity, maps compactly supported functions onto compactly supported functions, and is a norm-decreasing projection of $L^p(G)$ onto its subspace consisting of K-right-invariant elements, for $1\le p\le\infty$. A function on G is called K-bi-invariant, if $f(kgk')=f(g)$, for every $k,k'\in K$. A K-bi-invariant function may be identified with a function on $\mathfrak{X}$ which is invariant under the action of K. If K acts transitively on Ω, the boundary of $\mathfrak{X}$, then K-bi-invariant functions on G are constant on the sets $\mathfrak{B}_n=\{x:\ d(x,o)=n\}$ and therefore correspond to *radial functions on* $\mathfrak{X}$.

We shall assume from now on that *the action of* K *on* Ω *is transitive.* Under these conditions we have the following result.

(4.1) *LEMMA. The subspace of* $L^1(G)$ *consisting of K-bi-invariant functions is a commutative subalgebra of* $L^1(G)$ *under convolution.*

PROOF. Let u and v be K-bi-invariant elements of $L^1(G)$ and let $k,k'\in K$ then

$$u*v(k'gk)=\int_G u(k'gh)v(h^{-1}k^{-1})dh=u*v(g).$$

This shows that the space of K-bi-invariant elements is an algebra. To prove that this algebra is commutative, we show that $g^{-1}\in KgK$. Indeed $\mathbf{d}(go,o)=\mathbf{d}(o,g^{-1}o)$, therefore there exists $k\in K$, such that $go=kg^{-1}o$. This means that $g^{-1}kg^{-1}=k'\in K$, and therefore $g^{-1}=k'gk^{-1}$. It follows that, if u and v are K-bi-invariant,

$$u*v(g)=\int_G u(gh)v(h^{-1})dh=\int_G u(gh)v(h)dh$$

$$=\int_G u(h)v(g^{-1}h)dh=\int_G v(g^{-1}h)u(h^{-1})dh=v*u(g^{-1})=v*u(g). \blacksquare$$

Observe that, if f is a locally integrable function on G, we may define

$${}^Kf^K(g)=\int_K\int_K f(kgk')dkdk',$$

which is a K-bi-invariant function. We observe that the map $f\to {}^Kf^K$ defines a norm-decreasing projection on each of the spaces $L^p(G)$, for $1\le p\le\infty$.

A pair (G,K), where G is a locally compact group and K a compact subgroup, is called a *Gelfand pair* if the convolution algebra $L^1(K\backslash G/K)$ of integrable K-bi-invariant functions is commutative. We have thus proved that, under the hypothesis that K acts transitively on Ω, (G,K) is a Gelfand pair. It is not difficult to prove that the transitive action of K on Ω is also a necessary condition for (G,K) to be a Gelfand pair. We will now characterize the multiplicative linear functionals on the algebra $L^1(K\backslash G/K)$.

Observe that, if Φ is a continuous linear functional on $L^1(K\backslash G/K)$, then Φ is the restriction of a continuous linear functional on $L^1(G)$. Therefore there exists an essentially bounded measurable function ϕ such that, for every $u\in L^1(K\backslash G/K)$,

$$\Phi(u)=\int_G u(g)\phi(g)dg.$$

But

$$\Phi(u)=\int_G\int_K\int_K u(k'gk)\phi(g)dgdk'dk=\int_G u(g){}^K\phi^K(g)dg.$$

Therefore we may choose the function ϕ to be K-bi-invariant. In this case ϕ is necessarily continuous, because it is constant on the disjoint open sets KgK.

(4.2) *LEMMA. Let ϕ be a bounded, continuous K-bi-invariant function; then the functional*

$$\Phi(u)=\int_G u(g)\phi(g)dg$$

is multiplicative on $L^1(K\backslash G/K)$ if and only if, for every g and $g'\in G$

$$\phi(g)\phi(g')=\int_K \phi(gkg')dk. \qquad (1)$$

PROOF. Suppose that the functional equation (1) holds and let $u,v\in L^1(K\backslash G/K)$. Then

$$\Phi(u)\Phi(v)=\int_G\int_G u(g)v(g)\phi(g)\phi(h)dhdg$$

$$=\int_K\int_G\int_G u(g)v(h)\phi(gkh)dhdgdk=\int_G\int_G u(g)v(h)\phi(gh)dhdg$$

$$=\int_G\int_G u(g)v(g^{-1}h)\phi(h)dgdh=\int_G \phi(h)v*u(h^{-1})dh=\Phi(v*u).$$

(We have used the fact that a K-bi-invariant function has the same value at g and g^{-1}.) Conversely, if $\Phi(v*u)=\Phi(u)\Phi(v)$, then

$$\int_G\int_G v(h)u(gh^{-1})\phi(g)dgdh=\int_G\int_G v(h)u(g)\phi(h)\phi(g)dgdh,$$

which implies, for every $k\in K$,

$$\int_G\int_G v(h)u(gh^{-1}k^{-1})\phi(g)dgdh=\int_G\int_G v(h)u(g)\phi(gkh)dhdg$$

$$=\int_G\int_G v(h)u(g)\phi(h)\phi(g)dgdh.$$

Integrating over K, one obtains

$$\int_G\int_G\int_K \phi(gkh)v(h)u(g)dkdgdh=\int_G\int_G v(h)u(g)\phi(h)\phi(g)dgdh.$$

The function $\int_K \phi(gkh)dk$ is defined on the Cartesian product of the double-coset spaces $K\backslash G/K \times K\backslash G/K$, because it is separately K-bi-invariant as a function of g and as a function of h. Since v and u are arbitrary K-bi-invariant integrable functions we conclude that

$$\int_K \phi(gkh)dk=\phi(g)\phi(h). \blacksquare$$

(4.3) *LEMMA. Let* ϕ *be a bounded K-bi-invariant function satisfying the functional equation*

$$\int_K \phi(gkh)dk=\phi(g)\phi(h).$$

Let ψ *be the radial function on* $\mathfrak{X}$ *defined by* $\psi(x)=\phi(g)$ *with* $g\mathbf{o}=x$. *Then* ψ *is spherical in the sense of* (2.2).

PROOF. Since ϕ is K-bi-invariant, ψ is radial. We only have to show that $L\psi(x)=\mu\psi(x)$ for some $\mu\in\mathbb{C}$, and all $x\in\mathfrak{X}$. Let μ be the value of ψ on the nearest neighbors of $\mathbf{o}$. Let $g\mathbf{o}=x$, and let $y=h\mathbf{o}$ be a nearest neighbor of x. Observe that gKg^{-1} is the stabilizer of $x=g\mathbf{o}$, and that as k runs over gKg^{-1}, $kh\mathbf{o}$ runs over the nearest neighbors of x. Therefore,

$$L\psi(x)=\int_{gKg^{-1}}\phi(kh)dk=\int_K \phi(gkg^{-1}h)dk=\phi(g)\phi(g^{-1}h)=\mu\phi(g),$$

because $\mathbf{d}(g^{-1}h\mathbf{o},\mathbf{o})=\mathbf{d}(h\mathbf{o},g\mathbf{o})=\mathbf{d}(x,y)=1$. Since $\phi(g)=\psi(x)$, it follows that $L\psi(x)=\mu\psi(x)$. $\blacksquare$

With a slight abuse of language we shall call a K-bi-invariant function satisfying the functional equation (1) a *spherical function,* since, by the preceding result, it identifies uniquely a spherical function on $\mathfrak{X}$. Indeed it is not difficult to show that a spherical function on $\mathfrak{X}$ gives rise to a continuous K-bi-invariant function on G, satisfying the condition (1). Unbounded K-bi-invariant functions on G may be associated to multiplicative linear functionals on the commutative algebra of K-bi-invariant compactly supported functions. However, from now on we shall only encounter bounded spherical functions.

5. Spherical representations. In this section we shall identify the irreducible unitary representations of G which have a matrix coefficient which is a spherical function.

Recall that a unitary representation π of the group G is a homomorphism of G into the group of unitary operators on a Hilbert space $\mathcal{H}_\pi$ such that, for all $\xi,\eta\in\mathcal{H}_\pi$ the function $u(g)=(\pi(g)\xi,\eta)$ is continuous on G.

A unitary representation is said to be irreducible if there exists no nontrivial, closed subspace of $\mathcal{H}_\pi$ which is preserved by the action of all unitary operators $\pi(g), g\in G$. In other words π is irreducible if, for every nonzero $\xi\in\mathcal{H}_\pi$, the closed span of $\{\pi(g)\xi: g\in G\}$ is all of $\mathcal{H}_\pi$. We shall now characterize certain irreducible representation of G. Recall that our hypothesis on G is that its action on $\mathfrak{X}$ and on Ω is transitive. As before we let K denote the compact subgroup which fixes a vertex o of $\mathfrak{X}$.

(5.1) *DEFINITION. Let π be an irreducible unitary representation of* G*; then π is said to be spherical (with respect to* K*) if there exists a nonzero* K*-invariant vector, that is if there exists a nonzero vector $\xi_\pi\in\mathcal{H}_\pi$ such that*

$\pi(k)\xi_\pi=\xi_\pi$ *for every* $k\in K$.

We first show that the irreducibility of π implies that the space of K-invariant vectors is at most one-dimensional.

(5.2) *LEMMA. Let* π *be an irreducible unitary representation of* G *and let* $\mathcal{H}_\pi^K=\{\xi\in\mathcal{H}_\pi:\ \pi(k)\xi=\xi$ for all $k\in K\}$. *Then* $\dim \mathcal{H}_\pi^K\le 1$.

PROOF. Observe that if $f\in L^1(K\backslash G/K)$ then $\pi(f)\xi\in\mathcal{H}_\pi^K$, because

$$\pi(k)\pi(f)\xi=\int_G f(g)\pi(k)\pi(g)\xi dg=\int_G f(k^{-1}g)\pi(g)\xi dg=\pi(f)\xi.$$

This means that π defines a representation of the commutative involution algebra $L^1(K\backslash G/K)$ on the space $\mathcal{H}_\pi^K$. If $\mathcal{H}_\pi^K$ is more than one-dimensional, then this representation must be reducible. In other words there exists a nontrivial proper subspace $\mathcal{W}\subseteq\mathcal{H}_\pi^K$ which is invariant under $\pi(f)$ for $f\in L^1(K\backslash G/K)$.

Let P_K be the projection of $\mathcal{H}_\pi$ onto $\mathcal{H}_\pi^K$, then, for $\xi\in\mathcal{H}_\pi$,

$$P_K\xi=\int_K \pi(k)\xi dk.$$

Let $0\neq\eta\in\mathcal{H}_\pi^K$, and suppose that $\eta\perp\mathcal{W}$. We shall prove that, for every $f\in L^1(G)$, $\pi(f)\eta\perp\mathcal{W}$, which is a contradiction, because π is irreducible. Indeed let $\xi\in\mathcal{W}$. Since both ξ and η are K-invariant,

$$\begin{aligned}(\xi,\pi(f)\eta)&=\int_G f(g)(\pi(k)\xi,\pi(g)\pi(k')\eta)dg\\ &=\int_K\int_K\int_G f(g)(\xi,\pi(k^{-1}gk')\eta)dgdkdk'\\ &=\int_G {}^K\!f^K(g)(\xi,\pi(g)\eta)dg=(\xi,\pi({}^K\!f^K)\eta)=0,\end{aligned}$$

because the orthogonal complement of $\mathcal{W}$ in $\mathcal{H}_\pi^K$ is invariant under $\pi({}^K\!f^K)$. ∎

We can now show that if ξ_π is a K-invariant vector of the spherical representation π, and $\|\xi_\pi\|=1$, then $(\pi(g)\xi_\pi,\xi_\pi)$ is a spherical function.

(5.3) *THEOREM. Let π be a spherical representation; then there is one and only one positive-definite spherical function ϕ which is a matrix coefficient of π, that is $(\pi(g)\xi_\pi,\xi_\pi)=\phi(g)$, where ξ_π is a K-invariant vector of norm 1. Conversely, if ϕ is a positive-definite spherical function then ϕ is a matrix coefficient of a spherical representation. In particular different spherical functions are coefficients of inequivalent representations.*

PROOF. If π is a spherical representation and $\phi(g)=(\pi(g)\xi_\pi,\xi_\pi)$, then

$$\phi(kgk')=(\pi(g)\pi(k')\xi_\pi,\pi(k^{-1})\xi_\pi)=\phi(g).$$

Therefore ϕ is K-bi-invariant, and obviously bounded. Furthermore $\|\xi_\pi\|=1$, which implies $\phi(e)=1$. Let $g,h\in G$; then

$$\int_K(\pi(gkh)\xi_\pi,\xi_\pi)dk=\int_K(\pi(k)\pi(h)\xi_\pi,\pi(g^{-1})\xi_\pi)dk$$
$$=(P_K\pi(h)\xi_\pi,\pi(g^{-1})\xi_\pi).$$

Since $\mathcal{H}_\pi^K$ is one-dimensional and $P_K\pi(h)\xi_\pi\in\mathcal{H}_\pi^K$, $P_K\pi(h)\xi_\pi=\alpha(h)\xi_\pi$. Thus

$$\int_K\phi(gkh)dk=\int_K(\pi(gkh)\xi_\pi,\xi_\pi)dk=\alpha(h)(\pi(g)\xi_\pi,\xi_\pi)=\alpha(h)\phi(g).$$

Letting $g=1_G$, one obtains $\alpha(h)=(\pi(h)\xi_\pi,\xi_\pi)=\phi(h)$. Thus the function ϕ satisfies the condition (4.2 (1)), and is a spherical function. No other matrix coefficient $(\pi(g)\eta,\eta)=\psi(g)$ can be a spherical function, because the K-bi-invariance of ψ implies that η is K-invariant, and $1=\psi(1_G)=\|\eta\|^2$. Thus by (5.2) $\eta=c\xi_\pi$ with $|c|=1$ and $\psi(g)=\phi(g)$. Conversely let ϕ be a positive-definite spherical function on G. Then ϕ defines, by the Gelfand-Naimark-Segal construction [**D1**, Prop. 2.4.4], a unitary representation on the Hilbert space $\mathcal{H}$ generated by linear combination of left translates of ϕ, with inner product defined as follows: $(\xi,\eta)=\sum_{i,j}c_id_j\phi(g_ig_j^{-1})$, when $\xi=\sum c_i\phi(g_ig)$ and $\eta=\sum d_j\phi(g_jg)$. For

$\xi=\sum c_i\phi(g_i g)$, define $\pi(h)\xi=\sum c_i\phi(h^{-1}g_i g)$. Then π is unitary if the inner product is defined as above. With these definitions, ϕ is a cyclic vector of the representation and $\phi=\xi_\pi$ is K-invariant. It remains to prove that the representation π is irreducible, in other words that every nonzero vector is a cyclic vector. If $\xi=\sum c_i\phi(g_i g)$, then $P_K\xi=\sum c_i\int_K\phi(g_i kg)dk=$ $(\sum c_i\phi(g_i))\xi_\pi$. Since linear combinations of left translates of ϕ are dense, we conclude that, for every $\xi\in\mathcal{H}$, $P_K\xi$ is a constant multiple of ξ_π. On the other hand, if $\xi\neq0$ and $P_K\pi(g)\xi=0$ for every $g\in G$, then $\pi(g)\xi\perp\xi_\pi$ for every g, which implies $\xi=0$, because ξ_π is cyclic. Therefore, for some $g\in G$, $P_K\pi(g)\xi\neq0$. In order to show that ξ is cyclic it is enough to prove that $\pi(g)\xi$ is cyclic for some g. We may assume therefore that $P_K\xi\neq0$. We conclude that

$$P_K\xi=\int_K\pi(k)\xi dk=c(\xi)\xi_\pi$$

is a nonzero multiple of ξ_π and since $P_K\xi$ is a limit of convex combinations of vectors $\pi(k)\xi$, with $k\in K\subseteq G$, we conclude that ξ_π belongs to the closed linear space generated by $\pi(G)\xi$. Since ξ_π is cyclic so is ξ and we have proved that π is irreducible. ∎

It remains to characterize the spherical functions which are positive-definite.

Observe that if ϕ is spherical then $\phi(g^{-1})=\phi(g)$. Therefore if ϕ is positive-definite it is also real-valued. Let μ be the value of ϕ on the elements such that $\mathbf{d}(g\mathbf{o},\mathbf{o})=1$. Then if we think of ϕ as a function defined on the vertices of the tree $\mathfrak{X}$, we have that $L\phi(x)=\mu\phi(x)$ (2.3). Furthermore $|\mu|\leq1$. This means that positive-definite spherical functions are μ-eigenfunctions of L, for real μ, satisfying $-1\leq\mu\leq1$. Observe that $\mathrm{Aut}(\mathfrak{X})^+\cap G$ is a subgroup of index 2. Therefore $(-1)^{\mathbf{d}(g\mathbf{o},\mathbf{o})}$ is a K-bi-invariant character of G, which is of course a positive-definite spherical function. Likewise the function identically 1 is a positive-definite spherical function.

We shall prove that a spherical function associated with a real eigenvalue $-1<\mu<1$ is positive-definite.

Observe that $\mu=\mu(z)=(q^z+q^{1-z})(q+1)^{-1}$ satisfies $-1<\mu<1$, if and only if $z=1/2+it$, with $t\in\mathbb{R}$, or $\mathrm{Im}(z)=0$ and $0\le\mathrm{Re}(z)\le1$. Let $\mathcal{K}(\Omega)$ be the space of cylindrical functions. For $\xi\in\mathcal{K}(\Omega)$, and $z\in\mathbb{C}$, define $\pi_z(g)\xi(\omega)=P^z(g,\omega)\xi(g^{-1}\omega)$.

Observe that the identity

$$P(gh,\omega) = P(g,\omega)P(h,g^{-1}\omega), \qquad (1)$$

which follows directly from the definition of $P(g,\omega)$ as a Radon-Nikodym derivative, implies that $\pi_z(gh)\xi=\pi_z(g)(\pi_z(h)\xi)$. In addition, if $z=1/2+it$,

$$\int_\Omega|\pi_z(g)\xi(\omega)|^2d\nu=\int_\Omega P(g,\omega)|\xi(g^{-1}\omega)|^2d\nu=\int_\Omega|\xi(\omega)|^2d\nu.$$

Therefore, for $\mathrm{Re}(z)=1/2$, π_z is a unitary representation of G in the space $L^2(\Omega,\nu)$. Let $\mathbf{1}$ be the function identically equal to 1 on Ω. Then

$$\phi_z(g)=\int_\Omega P^{1/2+it}(g,\omega)d\nu=(\pi_z(g)\mathbf{1},\mathbf{1}).$$

Therefore ϕ_z is a diagonal coefficient of a unitary representation and so it is a positive-definite function. This means that the spherical functions associated to the eigenvalues $\mu=\mu(1/2+it)=(q+1)^{-1}q^{1/2}2\mathcal{R}e(q^{it})$ are all positive-definite. Observe that as t ranges over the reals $\mathcal{R}e(q^{it})$ assumes all values in the interval $[-1,1]$. This means that $\mu(1/2+it)$ assumes all the values in the interval $[-2q^{1/2}(q+1)^{-1},2q^{1/2}(q+1)^{-1}]$.

Next we consider the spherical functions associated to the real eigenvalues μ satisfying $q^{1/2}(q+1)^{-1}<\mu<1$, or $-1<\mu<-q^{1/2}(q+1)^{-1}$. Observe that in this case $\mu=\pm(q^s+q^{1-s})(q+1)^{-1}$, with $0<s<1$, and $s\ne1/2$. We may assume that $\mu>0$, because, if $L\phi_s=\mu(s)\phi_s$, $s'=s+i\pi/\ln q$, and $\phi_{s'}(g)=(-1)^{d(go,o)}\phi_s$, then $L\phi_{s'}=-\mu(s)\phi_{s'}=\mu(s')\phi_{s'}$.

Recall that the identity (1), the definition of $P(g,\omega)$ as a Radon-Nikodym derivative and the fact that $P(g,g\omega)=P^{-1}(g^{-1},\omega)$ imply

$$\phi_s(g^{-1}h)=\int_\Omega P^s(g^{-1}h,\omega)d\nu=\int_\Omega P^s(g^{-1},\omega)P^s(h,g\omega)d\nu$$

$$=\int_\Omega P^s(g^{-1},g^{-1}\omega)P^s(h,\omega)P(g,\omega)d\nu=\int_\Omega P^{1-s}(g,\omega)P^s(h,\omega)d\nu.$$

In other words, in terms of the intertwining operators I_s

$$\phi_s(g^{-1}h)=\int_\Omega I_sP^s(g,\omega)P^s(h,\omega)d\nu.$$

Therefore, if $\xi=\sum c_iP^s(g_i,\omega)$, then

$$\sum c_i\bar{c}_j\phi_s(g_i^{-1}g_j) = \int_\Omega (I_s\xi)\bar{\xi}d\nu.$$

It follows that in order to show that ϕ_s is positive-definite it suffices to show that, for every $\xi\in\mathcal{K}(\Omega)$,

$$\int_\Omega (I_s\xi)(\omega)\overline{\xi(\omega)}d\nu\geq 0.$$

By (3.2) I_s has a set of eigenvectors ξ_x which spans $\mathcal{K}(\Omega)$. If $0<s<1$, the corresponding eigenvalues are the positive numbers

$$(q^{1-s}-q^{s-1})(q^s-q^{-s})^{-1}q^{-(2s-1)d(x',o)}.$$

Therefore $(I_s\xi,\xi)\geq 0$, which proves that the spherical functions ϕ_s are positive-definite.

As we saw before the spherical functions ϕ_s , with $0<s<1$, define unitary irreducible representations of G with a K-invariant vector. We shall prove that the representation associated to ϕ_s is equivalent to a representation defined on a Hilbert space $\mathcal{H}_s$ which is the completion, under an appropriate norm, of the space of cylindrical functions $\mathcal{K}(\Omega)$.

We observe first that the sesquilinear form $(I_s\xi,\eta)=(\xi,\eta)_s$ defines an inner product on $K(\Omega)$. We let $\mathcal{H}_s$ be the completion of $\mathcal{K}(\Omega)$ with respect to this inner product. The space $\mathcal{K}(\Omega)$ is in one to one correspondence through the Poisson transform $\mathcal{P}_{1-s}$

with the space of linear combinations of left translations of ϕ_s and the relationship

$$\sum c_i \bar{c}_j \phi_s(g_i^{-1} g_j) = \int_\Omega (I_s \xi)(\omega) \overline{\xi(\omega)} d\nu = (\xi, \xi)_s ,$$

for $\xi = \sum c_i P^s(g_i, \omega)$, shows that the norm of $\mathcal{H}_s$ corresponds on $K(\Omega)$ to the norm defined by ϕ_s on the linear span of its left translates. Left translation on the latter space (which defines the unitary representation associated to ϕ_s) corresponds on $\mathcal{K}(\Omega)$ to the representation $\pi_s(g)\xi(\omega) = P^s(g,\omega)\xi(g^{-1}\omega)$.

As a last observation we notice that if $\xi \in \mathcal{K}_n(\Omega)$, and $\Delta_k \xi$ is the orthogonal projection of ξ on the space generated by $\{\xi_x : d(x,o)=k\}$, then $\xi = \sum_k \Delta_k \xi$. The functions $\Delta_k \xi$ are all orthogonal and are eigenfunctions of I_s with respect to the eigenvalue

$$c(k,s) = (q^{1-s} - q^{s-1})(q^s - q^{-s})^{-1} q^{-(k-1)(2s-1)},$$

for $k \geq 1$, and $c(0,s)=1$. Therefore

$$\|\xi\|^2_{\mathcal{H}} = (I_s \xi, \xi) = \sum_k c(k,s) \|\Delta_k \xi\|^2_{L^2(\Omega)} .$$

We should observe that $\sum_{k=0}^{n} \Delta_k \xi$ is nothing but the orthogonal projection of ξ into the subspace spanned by the characteristic functions of the sets $\Omega(x)$ with $|x|=n$.

(5.4) *DEFINITION. The series of unitary representations* $\{\pi_{1/2+it} : t \in \mathbb{R}\}$ *is called the unitary principal series of representations of G. The series of representations*

$$\{\pi_s : s \neq 1/2 \ 0 \leq s \leq 1\} \cup \{\pi_{s+i\pi/\ln q} : s \neq 1/2 \ 0 \leq s \leq 1\}$$

is called the unitary complementary series.

6. The resolvent of the Laplace operator and the spherical Plancherel formula. In this section we compute the resolvent of the Laplace operator L, acting on the Hilbert space $\ell^2(\mathfrak{X})$.

Define, for $f\in\ell^2(\mathfrak{X})$ and $n>0$,

$$L_n f(x) = (q+1)^{-1}q^{-n+1}\sum_{d(x,y)=n} f(y) .$$

Observe that $L=L_1$. For completeness, we may define $L_0=I$, the identity operator.

We shall prove that if $\mathcal{R}ez>1/2$, and $\mu(z)=(q^z+q^{1-z})(q+1)^{-1}$, then the series

$$H_z = L_0+(q+1)q^{-1}\sum_{n=1}^{\infty} q^{n(1-z)} L_n$$

converges absolutely in the operator norm, and that

$$(L-\mu(z))^{-1} = (q+1)(q^{-z}-q^{z})^{-1}H_z .$$

Let $\mathbf{o}$ be a fixed vertex and define $\mathfrak{B}_m=\{x: \mathbf{d}(\mathbf{o},x)=m\}$. We denote by χ_m the characteristic function of the set $\mathfrak{B}_m$.

(6.1) *LEMMA. Let* f *be a function supported on* $\mathfrak{B}_k$*; then for any positive integers* m *and* n*,*

$$\|\chi_m(L_n f)\|_2\le (q^{-n+1}/(q+1))^{1/2}\ \|f\|_2 .$$

Moreover, if $\chi_m(L_n f)\neq 0$*, then* $n+k-m$ *is even and* $|n-k|\le m\le n+k$.

PROOF. If, for some $x\in\mathfrak{B}_m$, we have $L_n f(x)\neq 0$, then there exists $t\in\mathfrak{B}_k$ such that $\mathbf{d}(t,x)=n$. Let y be such that $[\mathbf{o},t]\cap[\mathbf{o},x]=[\mathbf{o},y]$. Then $n+k-\mathbf{d}(\mathbf{o},x)=\mathbf{d}(t,x)+\mathbf{d}(t,\mathbf{o})-\mathbf{d}(\mathbf{o},x)=2\mathbf{d}(t,y)$. Moreover $m=\mathbf{d}(\mathbf{o},x)$ and the triangle inequality implies $|\mathbf{d}(t,x)-\mathbf{d}(t,\mathbf{o})|\le \mathbf{d}(\mathbf{o},x)\le \mathbf{d}(t,x)+\mathbf{d}(t,\mathbf{o})$. This proves the second part of the lemma. For the first part we first assume that $m=n+k$. Let $\mathbf{d}(\mathbf{o},x)=m$. The assumption implies that there is only one element $y_x\in\mathfrak{B}_k$ such that $\mathbf{d}(y_x,x)=n$. Hence $(L_n f)(x)=q^{-n+1}(q+1)^{-1}f(y_x)$. On the other hand, for each $y\in\mathfrak{B}_k$ there are q^n elements x such that $\mathbf{d}(\mathbf{o},x)=m$, and $y\in[\mathbf{o},x]$. Therefore,

$$\|\chi_m(L_n f)\|_2^2 \ =\sum_{d(o,x)=m}|L_n f(x)|^2 \ = \ (q^{-n+1}/(q+1))^2q^n\sum_{d(o,y)=k}|f(y)|^2$$

$$\le (q/(q+1))^2q^{-n}\|f\|_2^2\le (q^{-n+1}/(q+1))\|f\|_2^2.$$

Suppose now that $m=n+k-2j$, with $j>0$. We define $F(y)=0$ if $\mathbf{d}(y,\mathbf{o})\neq k-j$,

$$F(y)=\Big[\sum_{d(y,t)=j}|f(t)|^2\Big]^{1/2},\ \text{if } d(y,o)=k-j.$$

Then $\|F\|_2=\|f\|_2$, and F is supported in $\mathfrak{B}_{k-j}$. Therefore by the first part of the proof $\|\chi_m(L_{n-j}F)\|_2\le(q/(q+1))^{1/2}q^{-n/2}\|F\|_2$. Let $d(o,x)=m$ and let y_x be the unique vertex satisfying $y_x\in[o,x]$ and $y_x\in[x,t]$ for some $t\in\mathfrak{B}_k$. Then $d(y_x,o)=k-j$ and $L_nf(x)=q^{-n+1}/(q+1)\sum_{d(y_x,t)=j}f(t)$. Thus, by the Cauchy-Schwarz inequality,

$$|L_nf(x)|\le q^{-n+1}(q+1)^{-1}q^{j/2}\Big[\sum_{d(y,t)=j}|f(t)|^2\Big]^{1/2}$$
$$= q^{-n+1}(q+1)^{-1}q^{j/2}F(y_x).$$

Since there is only one element y_x in the support of F and at distance n-j from x, $L_{n-j}F(x)=(q/(q+1))q^{-n+j}F(y_x)$. Therefore $|L_nf(x)|\le q^{-j/2}|L_{n-j}F(x)|$. We conclude that $\|\chi_m(L_nf)(x)\|_2^2 \le q^{-n+1}(q+1)\|F\|_2^2$. The proof is completed by observing that $\|F\|=\|f\|$. ∎

(6.2) *LEMMA. Let $f\in\ell^2(\mathfrak{X})$; then*

$$\|L_nf\|_2^2\le q^{1-n}(q+1)(n+1)\|f\|_2.$$

PROOF. Let $f_k=\chi_kf$, then $f=\sum f_k$ and $\|f_k\|_2^2=\sum\|f_k\|_2^2$. We have $\|\chi_m(L_nf)\|_2\le\sum_k\|\chi_m(L_nf_k)\|_2$. But $\chi_m(L_nf_k)\ne0$ only if $|m-n|\le k\le m+n$, in other words only if $k=\min(n,m)-2j$, with $j=0,\ldots,m+n$. It follows that

$$\|\chi_m(L_nf)\|_2\le\sum_{j=0}^{\min(n,m)}\|\chi_m(L_nf_{m+n-2j})\|_2$$
$$\le q^{(1-n)/2}(q+1)^{-1}(n+1)^{1/2}\Big[\sum_{j=0}^{\min(n,m)}\|f_{m+n-2j}\|_2^2\Big]^{1/2}.$$

Finally, $\|L_nf\|_2^2=\sum_m\|\chi_m(L_nf)\|_2^2\le q^{1-n}(q+1)^{-1}(n+1)\sum_{j=0}^{n}\sum_{k=n-j}^{\infty}\|f\|_2^2\le q^{n-1}(q+1)^{-1}(n+1)^2\|f\|_2^2$. ∎

The previous lemmas imply that the operator

$$H_z = L_0+(q+1)q^{-1}\sum_{n=1}^{\infty} q^{n(1-z)}L_n$$

is well defined by the absolute convergence of the series, and it is bounded. Note that $H_z\delta_x(y)=q^{-zd(x,y)}$. This implies that, for $y\neq x$,

$$L(H_z\delta_x)(y)= (q+1)^{-1}(q^{-z(d(x,y)-1)}+qq^{-z(d(x,y)+1)})$$

$$= \mu(z)(H_z\delta_x)(y).$$

In other words the operator $(L-\mu(z))H_z$ applied to δ_x yields a multiple of δ_x and indeed $((L-\mu(z))H_z\delta_x,\delta_x)= q^{-z}-\mu(z) = (q^{-z}-q^{z})/(q+1)$. It follows that

$$(L-\mu(z))^{-1} = (q+1)(q^{-z}-q^{z})^{-1}H_z.$$

We summarize this result in a proposition.

(6.3) *PROPOSITION. If* $\mathcal{R}ez>1/2$, *and* $\mu(z)=(q^{z}+q^{1-z})/(q+1)$, *the operator* $L-\mu(z)$ *is invertible in* $\ell^2(\mathfrak{X})$ *and*

$$(L-\mu)^{-1}= (q+1)(q^{-z}-q^{z})^{-1}H_z.$$

The spectrum of L *in* $\ell^2(\mathfrak{X})$ *is the interval* $[-\sqrt{q/(q+1)},\sqrt{q/(q+1)}]$.

PROOF. The first part of the proposition is proved in the foregoing remarks. Since every μ which does not belong to the interval above satisfies $\mu=\mu(z)$, with $\mathcal{R}ez>1/2$, we also have that the spectrum of L is contained in the interval $[-\sqrt{q(q+1)},\sqrt{q(q+1)}]$. Suppose now that μ is a number in this interval which does not belong to the spectrum. Then $\mu=\mu(1/2+it)$ for some real t. For $\varepsilon>0$,

$$(L-\mu(1/2+\varepsilon+it))^{-1}\delta_x(y)$$

$$= (q+1)(q^{1/2+\varepsilon+it}-q^{-1/2-\varepsilon-it})^{-1}q^{(1/2+\varepsilon+it)d(x,y)}.$$

As $\varepsilon\to0$ the constant which multiplies the function $q^{(1/2+\varepsilon+it)d(x,y)}$ is bounded away from zero. But, for fixed x, the norm in ℓ^2 of the same function becomes unbounded, contradicting the fact that the resolvent is continuous in its domain of definition. ∎

It is easy to see, using the expression for H_z, that $(\mu(z)-L)^{-1}(\delta_o)(x)=(q+1)(q^z-q^{-z})^{-1}q^{-d(x,o)z}$.

Let $r_\mu(x)=(\mu-L)^{-1}(\delta_o)(x)$. We prove now the spherical Plancherel formula of the tree. We recall the Carlemann formula [DS, p. 920]: if T is a bounded self-adjoint operator on a Hilbert space $\mathcal{H}$ and [a, b] is the spectrum of T then for every $\xi,\eta\in\mathcal{H}$ the following equality holds:

$$(\xi,\eta)=-\lim_{\delta\to0^+}\lim_{\varepsilon\to0^+}(1/2\pi i)\int_{a+\delta}^{b-\delta}((R_{\mu+i\varepsilon}-R_{\mu-i\varepsilon})\xi,\eta)d\mu$$

where $R_\mu=(\mu-T)^{-1}$.

In particular if T=L, $\mathcal{H}=\ell^2(\mathfrak{X})$, $\xi=\delta_o$ and $\eta=\bar{f}$ is a function on $\mathfrak{X}$ with finite support, then we obtain

$$f(o)=-\lim_{\delta\to0^+}\lim_{\varepsilon\to0^+}(1/2\pi i)\int_{-\sqrt{q(q+1)}+\delta}^{\sqrt{q(q+1)}-\delta}(r_{\mu+i\varepsilon}-r_{\mu-i\varepsilon},\bar{f})d\mu$$

$$=\sum_x f(x)\left[-\lim_{\delta\to0^+}\lim_{\varepsilon\to0^+}(1/2\pi i)\int_{-\sqrt{q(q+1)}+\delta}^{\sqrt{q(q+1)}-\delta}(r_{\mu+i\varepsilon}-r_{\mu-i\varepsilon})(x)d\mu\right].$$

We observe that if $\varepsilon\to0^+$ and $\mu=\mu(1/2+it)$ then $\mu+i\varepsilon=\mu(z_\varepsilon)$ with $\mathcal{R}e z_\varepsilon>1/2$ and $z_\varepsilon\to1/2+it$ for $\varepsilon\to0^+$. Therefore

$$\lim_{\varepsilon\to0^+}r_{\mu+i\varepsilon}(x)=(q+1)(q^{1/2+it}-q^{-1/2-it})^{-1}q^{-d(x,o)(1/2+it)}.$$

On the other hand $\mu-i\varepsilon=\mu(w_\varepsilon)$ with $\mathcal{R}e w_\varepsilon>1/2$ and $w_\varepsilon\to1/2-it$. Therefore

$$\lim_{\varepsilon\to0^+}r_{\mu-i\varepsilon}=(q+1)(q^{1/2-it}-q^{-1/2+it})^{-1}q^{-d(x,o)(1/2-it)}.$$

For $\mu=\mu(1/2+it)$ $0<t<\pi/\ln q$ we have

$$f(o)=\sum_x f(x)\left[-\lim_{\delta\to0^+}(1/2\pi i)\int_{-\sqrt{q(q+1)}+\delta}^{\sqrt{q(q+1)}-\delta}\lim_{\varepsilon\to0^+}(r_{\mu+i\varepsilon}-r_{\mu-i\varepsilon})(x)d\mu\right]$$

$$=\sum_x f(x)\left[\lim_{\delta\to 0^+}(1/2\pi i)\int_\delta^{\pi/\ln q-\delta}((q+1)(q^{1/2+it}-q^{-1/2-it})^{-1}q^{-d(x,o)(1/2+it)}-\right.$$

$$\left.-(q+1)(q^{1/2-it}-q^{-1/2+it})^{-1}q^{-d(x,o)(1/2-it)})\mu'(1/2+it)dt\right]$$

$$=\sum_x f(x)\left[(\ln q/2\pi)\int_0^{\pi/\ln q}(A(t)q^{-d(x,o)(1/2+it)}+B(t)q^{-d(x,o)(1/2-it)})dt\right]$$

where $A(t)=(q^{1/2+it}-q^{1/2-it})/(q^{1/2+it}-q^{-1/2-it})$ and

$$B(t)=(q^{1/2-it}-q^{1/2+it})/(q^{1/2-it}-q^{-1/2+it}).$$

We have that

$$A(t)q^{-d(x,o)(1/2+it)}+B(t)q^{-d(x,o)(1/2-it)}$$
$$=(q/(q+1))\left[c(1/2-it)^{-1}q^{-d(x,o)(1/2+it)}+c(1/2+it)^{-1}q^{-d(x,o)(1/2-it)}\right]$$
$$=(q/(q+1))|c(1/2+it)|^{-2}\phi_{1/2+it}(x).$$

We recall that $c(1/2+it)=\overline{c(1/2-it)}$ (see Proposition (2.4)). It follows that

$$f(o)=\sum_x f(x)(q\ln q)(2\pi(q+1))^{-1}\int_0^{\pi/\ln q}\phi_{1/2+it}(x)|c(1/2+it)|^{-2}dt$$
$$=(q\ln q)(2\pi(q+1))^{-1}\int_0^{\pi/\ln q}(\phi_{1/2+it},\bar{f})|c(1/2+it)|^{-2}dt.$$

Therefore we have proved the following theorem.

(6.4) *THEOREM. Let* J *be the interval* $[0,\pi/\ln q]$ *and* $dm(t)$ *be the measure on* J *defined as follows:*

$$dm(t)=(q\ln q)(2\pi(q+1))^{-1}|c(1/2+it)|^{-2}dt.$$

Then for every f *on* $\mathfrak{X}$ *with finite support*

$$f(o)=\int_J(f,\phi_{1/2+it})dm(t).$$

Let Γ be a faithful transitive subgroup of $\mathrm{Aut}(\mathfrak{X})$; then Theorem (6.4) implies that, for $f=g^{*}*g$,

$$\|g\|_2^2=\int_J (g^{*}*g,\phi_{1/2+it})dm(t)=\int_J \|\pi_{1/2+it}(g)1\|^2_{L^2(\Omega)} dm(t)$$

for every function g on Γ with finite support. Also, Theorem (6.4) implies that

$$\int_J \phi_{1/2+it}(x)dm(t)=\delta_x(o).$$

We deduce from Theorem (6.4) the spherical Plancherel formula in the sense of Gelfand pairs [F, Th. IV, 2] for any closed subgroup of $\mathrm{Aut}(\mathfrak{X})$ acting transitively on $\mathfrak{X}$ and Ω (see Section 4).

(6.5) *COROLLARY. Let* G *be a closed subgroup of* $\mathrm{Aut}(\mathfrak{X})$ *acting transitively on* $\mathfrak{X}$ *and* Ω. *Fix the Haar measure on* G *so that* K *has measure equal to* 1. *Let* J *and* dm(t) *be as in Theorem* (6.4). *Then for every* K-*left-invariant function* f *on* G *with compact support we have*

$$\|f\|_2^2=\int_J \|\pi_{1/2+it}(f)1\|^2_{L^2(\Omega)} dm(t)=\int_J \|\pi_{1/2+it}(f)\|_{HS}^2 dm(t).$$

PROOF. By [F, Th. IV, 2] it is enough to prove the equality for every K-bi-invariant function f with compact support. Theorem (6.4) implies that $\int_J \phi_{1/2+it}(g)dm(t)=\chi_K(g)$. We have $\pi_{1/2+it}(f)=\pi_{1/2+it}(f)\pi_{1/2+it}(k)$ for every $k\in K$. This implies that $\pi_{1/2+it}(f)=\pi_{1/2+it}(f)P_o$ where P_o is the one-dimensional orthogonal projection onto the space of constant functions on Ω. Therefore $\|\pi_{1/2+it}(f)\|_{HS}=\|\pi_{1/2+it}(f)1\|_{L^2(\Omega)}$. One obtains

$$\|\pi_{1/2+it}(f)1\|^2_{L^2(\Omega)}$$

$$= (\pi_{1/2+it}(f^* * f)1, 1) = \int_\Omega \int_G f^* * f(g)\pi_{1/2+it}(g)1 \, dg d\omega$$

$$= \int_G f^* * f(g)\phi_{1/2+it}(g) \, dg.$$

Since $f^* * f$ is K-bi-invariant, it follows that

$$\int_J \|\pi_{1/2+it}(f)1\|^2_{L^2(\Omega)} dm(t) = \int_G f^* * f(g) \int_J \phi_{1/2+it}(g) \, dm(t) dg$$

$$= \int_G f^* * f(g)\chi_K(g) dg = f^* * f(1_G) = \|f^*\|_2^2 = \|f\|_2^2. \blacksquare$$

7. The restriction problem. In the previous sections we saw that, if G is a closed subgroup of Aut($\mathfrak{X}$) which acts transitively on $\mathfrak{X}$ and its boundary, then (G,K) is a Gelfand pair, where $K=\{g\in G: go=o\}$. This implies the irreducibility of spherical representations. Recall that, if $\mathcal{R}ez=1/2$ or $\mathcal{I}mz=0$ and $0<\mathcal{R}ez<1$, the spherical unitary representations may be defined as $\pi_z(g)\xi(\omega)=P^z(g,\omega)\xi(g^{-1}\omega)$, where ξ is an element of a suitable Hilbert space completion of $\mathcal{K}(\Omega)$. For the definition of π_z there is no need to assume that G acts transitively on the boundary. We may for instance define spherical representation for a finitely generated free group acting on its Cayley graph. This was done in **[F-T P2]**, where it is also proved that spherical representations of such a free group are irreducible. A much stronger result is in fact true.

(7.1) *THEOREM. Let* Γ *be a discrete subgroup of* Aut($\mathfrak{X}$) *with the property that* Aut($\mathfrak{X}$)/Γ *is compact. Let* $-1<\mu(z)<1$, *with* $\mu(z) =(q^z+q^{1-z})(q+1)^{-1}$. *Let* π_z *be the corresponding spherical representation. Then*

(1) *if* $\mu(z)\neq 0$ *then* π_z *restricted to* Γ *is irreducible.*

(2) *if* $\mu(z)=0$ *and* $\Gamma\not\subset \mathrm{Aut}(\mathfrak{X})^+$ *then* π_z *restricted to* Γ *is irreducible.*

(3) *if* $\mu(z)=0$ *and* $\Gamma\subseteq \mathrm{Aut}(\mathfrak{X})^+$ *then* π_z *restricted to* Γ *is the sum of exactly two irreducible subrepresentations.*

The complete proof of this theorem is contained in **[St2]**. In these notes we shall give a proof of this result under a number of simplifying assumptions which allow us to avoid certain technical details. This simpified version will contain however the main ideas of the proof. A discussion of how the proof may be completed to yield the full strength of the theorem is contained, in part, in the section of notes and remarks to this chapter. Our purpose here is to prove the following version of the theorem.

(7.2) *THEOREM.* *Let* Γ *be a discrete subgroup of* $\mathrm{Aut}(\mathfrak{X})$. *Suppose that* $\{g\in\Gamma:\ gx=x\}=\{e\}$, *for every* x. *Suppose further that* $\mathrm{Aut}(\mathfrak{X})/\Gamma$ *is compact, or, what is the same, that* Γ *acting on* $\mathfrak{X}$ *has finitely many orbits. Let* $q^{2it}\neq\pm 1$, *and let* $\pi=\pi_{1/2+it}$. *Then the restriction of* π *to* Γ *is irreducible.*

Comparison between the hypotheses of (7.1) and (7.2) shows that (7.2) consists of part (1) of (7.1) for representations of the principal series, under the assumption that no element of Γ except the identity leaves a vertex fixed.

We write $\pi(g)\xi=P^{1/2+it}(g,\omega)\xi(g^{-1}\omega)$, with $\xi\in L^2(\Omega,\nu)$, where ν is the isotropic measure defined with reference to a fixed vertex $o\in\mathfrak{X}$. In this context the function **1**, which is identically 1 on Ω, is the K-invariant vector.

Let π_Γ be the restriction to Γ of π. We must prove that every vector $\xi\in L^2(\Omega)$ is a cyclic vector for π_Γ. The proof is carried out in three steps.

STEP A. *Construct bounded operators* P_ε *in the* C^**-algebra of* π_Γ *with* $\|P_\varepsilon\|=O(1)$.
STEP B. *Show that, for each* $g_1,g_2\in\Gamma$, $(P_\varepsilon\pi(g_1)\mathbf{1},\pi(g_2)\mathbf{1})$ *converges to* $(P_0\pi(g_1)\mathbf{1},\pi(g_2)\mathbf{1})$, *where* P_0 *is the orthogonal projection onto the space of constant multiples of* $\mathbf{1}$.
STEP C. *Show that* $\mathbf{1}$ *is a cyclic vector for* π_Γ.

We prove now that these three steps are sufficient for the proof of the theorem. Indeed, by STEP C, if ξ is a nonzero vector there exists $\gamma\in\Gamma$, such that $(\pi(\gamma)\xi,\mathbf{1})\neq0$. Therefore $P_0\pi(\gamma)\xi\neq0$. But, by STEP A and STEP B, P_0 may be approximated in the weak operator topology by elements in the span of $\pi(\Gamma)$. This means that $P_0\pi(\gamma)\xi$, which is a nonzero multiple of $\mathbf{1}$, is in the weakly closed linear span of $\pi(\Gamma)\xi$. But STEP C says that $\mathbf{1}$ is a cyclic vector and therefore every vector is in the closed linear span of $\pi(\Gamma)\xi$.

We conclude this section with a remark on the spherical representation of the principal series which is used later.

Given two unitary representations π_1 and π_2 of a locally compact group G, we say that the first is weakly contained in the other if, for every $f\in L^1(G)$, $\|\pi_1(f)\|\le\|\pi_2(f)\|$. We know from **[D1]** that, if π_1 is a cyclic representation with cyclic vector ξ, then π_1 is weakly contained in π_2 whenever the positive-definite function $(\pi_1(g)\xi,\xi)$ may be approximated uniformly on compact sets by positive-definite coefficients of the representations π_2.

(7.3) *LEMMA. Every representation* $\pi_{1/2+it}$ *of the principal series of* $\mathrm{Aut}(\mathfrak{X})$ *is weakly contained in the regular representation.*

PROOF. By the foregoing remarks it suffices to show that the spherical function $\phi_{1/2+it}$ may be approximated on compact sets by positive-definite coefficients of the regular representation. By (2.4) $\phi_{1/2+it}\in L^{2+\varepsilon}$, for every $\varepsilon>0$. On the other hand, if $1/2<s<1$, $\lim_s \phi_s(g)=1$, uniformly on compact sets, and $\phi_s \in L^{1/(1-s)}(\mathrm{Aut}(\mathfrak{X}))$. Therefore $\lim_{s\to 1} \phi_s\phi_{1/2+it} = \phi_{1/2+it}$ uniformly on compact sets, and $\phi_s\phi_{1/2+it} \in L^2$. The latter condition implies **[D1**, Th. 13.8.6] that $\phi_s\phi_{1/2+it}$ is a positive-definite coefficient of the regular representation. ∎

8. Construction and boundedness of P_ε (STEP A). Throughout this section, for notational convenience, let $G=\mathrm{Aut}(\mathfrak{X})$. Let $\varepsilon>0$, and $z=z(\varepsilon)=1/2+\varepsilon+it$. Let $|G/\Gamma|$ be the volume of G/Γ, or, what is the same, the number of orbits of Γ in $\mathfrak{X}$. Define $h_z(g)= q^{-zd(go,o)}$, and

$$P_\varepsilon=|G/\Gamma|\ \pi_\Gamma(\mathcal{R}e(D(\varepsilon)h_z))=|G/\Gamma|\sum_{\gamma\in\Gamma} \mathcal{R}e[D(\varepsilon)h_z(\gamma)]\ \pi(\gamma)$$

where

$$D(\varepsilon)^{-1}=\int_\Omega\int_G h_z(g)P^{1/2+it}(g,\omega)\ dg d\nu = (\pi(h_z)\mathbf{1},\mathbf{1}).$$

Observe that, since h_z is K-bi-invariant, the integral above may be written as a sum:

$$\sum_{x\in\mathfrak{X}} h_z(x) \int_\Omega P^{1/2+it}(x,\omega)\ d\nu = \sum_{x\in\mathfrak{X}} h_z(x)\phi(x).$$

But

$$\phi(x)=c(1/2+it)q^{-(1/2+it)d(o,x)}+c(1/2-it)q^{-(1/2-it)d(o,x)}.$$

Therefore,

$$D(\varepsilon)^{-1}=(q+1)q^{-1}\left[c(1/2+it)\sum_{n=0}^{\infty} q^{-n(\varepsilon+2it)}+c(1/2-it)\sum_{n=0}^{\infty} q^{-\varepsilon n}\right]$$

$$= (q+1)q^{-1}\left[c(1/2+it)\ (1-q^{-(\varepsilon+2it)})^{-1}+ c(1/2-it)(1-q^{-\varepsilon})^{-1}\right].$$

This implies that $\varepsilon/D(\varepsilon)$ is bounded, as well as bounded away from zero as $\varepsilon\to 0$. Notice that $\sum_{\gamma\in\Gamma}|h_z(\gamma)|<\infty$. Therefore $\pi_\Gamma(h_z)$ belongs to the C^*-algebra of π_Γ, and so does P_ε.

Our next task is to show, according to STEP A, that $\|P_\varepsilon\|=O(1)$. Because of what we know of $D(\varepsilon)$ it suffices to show that $\|\pi_\Gamma(h_z)\|=O(\varepsilon)$. Let ρ_Γ be the right regular representation of Γ. We observe that π_Γ is weakly contained in ρ_Γ. This follows from (7.3). Therefore it suffices to show that $\|\rho_\Gamma(h_z)\|=O(\varepsilon)$. The function $h_z(\gamma)$ acting by right convolution on $\ell^2(\Gamma)$ may be thought of as an operator H_z^Γ defined on the subspace $\ell^2(\Gamma o)\subseteq\ell^2(\mathfrak{X})$ and having matrix coefficients $(H_z^\Gamma\delta_{\gamma o},\delta_{\gamma' o})=h_z(\gamma^{-1}\gamma')$. Let Q be the orthogonal projection of $\ell^2(\mathfrak{X})$ onto $\ell^2(\Gamma o)$. Let H_z be the operator on $\ell^2(\mathfrak{X})$ having matrix coefficients $(H_z\delta_x,\delta_y)=q^{-zd(x,y)}$. Then H_z^Γ is the restriction to $\ell^2(\Gamma o)$ of QH_zQ^{-1}. Therefore, in order to estimate the norm of H_z^Γ it suffices to estimate the norm of H_z. Recall that, according to (6.3),

$$(L-\mu(z))^{-1}= (q+1)(q^z-q^{-z})^{-1}H_z .$$

Our hypotheses imply that the coefficient multiplying H_z in the equation above is bounded away from zero as $\varepsilon\to 0$. Therefore in order to estimate the norm of $\pi_\Gamma(h_z)$ it suffices to estimate the norm of the resovent $(L-\mu(z))^{-1}$ of the Laplace operator on the tree $\mathfrak{X}$, where $\mu(z)=(q+1)(q^z+q^{1-z})$, $z=1/2+\varepsilon+it$, and ε is a small positive number. We shall use now a classical estimate for the resolvent of a self-adjoint operator.

(8.1) *LEMMA.* *Let* L *be a self-adjoint operator on a Hilbert space, and* μ *a complex number. Let* $d(\mu)$ *be the distance of* μ *from the convex hull of* L. *Then*

$$\|(L-\mu)^{-1}\|\le d(\mu)^{-1}.$$

A reference for the proof of the Lemma is [T, Th.6.2.A].

(8.2) *COROLLARY.* For $z=1/2+\varepsilon+it$, *and under the hypothesis of* (7.2), $\|\pi_\Gamma(h_z)\|=O(1/\varepsilon)$, *and* $\|P_\varepsilon\|=O(1)$, *as* $\varepsilon\to 0$.

PROOF. The first estimate follows from the lemma above and the remarks preceding it, as soon as we observe that $d(\mu)=\mathcal{I}m(\mu(z))\simeq\varepsilon$. The estimate for P_ε follows from the fact that $P_\varepsilon=|G/\Gamma|\pi_\Gamma(\mathcal{R}e(D(\varepsilon)h_z)$, and $D(\varepsilon)\simeq\varepsilon$. ∎

9. Approximating the projection P_0 (STEP B). Having completed STEP A of the proof in the previous section, our next task is to show that, if $g_1, g_2 \in \Gamma$,

$$\lim_{\varepsilon\to 0} (P_\varepsilon \pi(g_1)\mathbf{1}, \pi(g_2)\mathbf{1}) = (P_0 \pi(g_1)\mathbf{1}, \pi(g_2)\mathbf{1}).$$

Recall that $P_\varepsilon = \pi_\Gamma(\mathcal{R}e(Dh_z))$. Therefore

$$(P_\varepsilon \pi(g_1)\mathbf{1}, \pi(g_2)\mathbf{1}) = |G/\Gamma| \sum_{\gamma\in\Gamma} \mathcal{R}eDh_z(\gamma)(\pi(\gamma g_1)\mathbf{1}, \pi(g_2)\mathbf{1})$$

$$= |G/\Gamma| \sum_{\gamma\in\Gamma} \mathcal{R}eDh_z(\gamma)\ \phi(g_2^{-1}\gamma g_1) = |G/\Gamma| \sum_{\gamma\in\Gamma} \mathcal{R}eDh_z(g_2\gamma)\phi(\gamma g_1).$$

We also observe that P_0 may be written as

$$P_0 = \pi(\mathcal{R}eDh_z) = \int_G \mathcal{R}eDh_z(g)\ \pi(g)\ dg.$$

Indeed $\pi(\mathcal{R}eDh_z)$ is a self-adjoint bounded operator which maps every vector into a constant multiple of $\mathbf{1}$ (this is because h_z is K-bi-invariant). But $D=D(\varepsilon)$ was chosen so that $(\pi(\mathcal{R}eDh_z)\mathbf{1},\mathbf{1})=1$. Therefore $\pi(\mathcal{R}eDh_z)$ must be the orthogonal projection onto the space of multiples of $\mathbf{1}$. We write then

$$(P_0\pi(g_1)\mathbf{1}, \pi(g_2)\mathbf{1}) = (\pi(\mathcal{R}eDh_z)\pi(g_1)\mathbf{1}, \pi(g_2)\mathbf{1})$$

$$= \int_G \mathcal{R}e(Dh_z)(g)\phi(g_2^{-1}gg_1)\ dg = \int_G \mathcal{R}e(Dh_z)(g_2 g)\phi(gg_1)\ dg.$$

We observe that $\lim_{\varepsilon\to 0} D(\varepsilon)=0$. It suffices therefore to prove that

$$|G/\Gamma| \sum_{\gamma\in\Gamma} \mathcal{R}eh_z(g_2\gamma)\phi(\gamma g_1) - \int_G \mathcal{R}eh_z(g_2 g)\phi(gg_1)dg = O(1). \tag{1}$$

To prove (1) it suffices to prove the assertion when $\mathcal{R}eh_z$ is replaced by h_z. We are thus led to consider the function $h_z(g_1 g)\phi(gg_2)$ as $\varepsilon\to 0$. Since all the functions appearing in (1) are uniformly bounded for $\varepsilon\to 0$, it suffices to prove (1) when g and γ are restricted outside a given compact subset of G. Furthermore the expression of ϕ as a linear combination of $h_{1/2+it}$ and $h_{1/2-it}$ makes (1) true as soon as we establish that the quantities

$$|G/\Gamma|\sum_{\gamma\in\Gamma} h_z(g_1\gamma)h_{1/2-it}(\gamma g_2)-\int_G h_z(g_1g)h_{1/2-it}(gg_2)dg \qquad (2)$$

and

$$|G/\Gamma|\sum_{\gamma\in\Gamma} h_z(g_1\gamma)h_{1/2+it}(\gamma g_2)-\int_G h_z(g_1g)h_{1/2+it}(gg_2)dg \qquad (3)$$

are uniformly bounded.

We shall find first a convenient expression for the functions $h_z(g_1g)h_{1/2+it}(gg_2)$ and $h_z(g_1g)h_{1/2-it}(gg_2)$ outside a compact subset of G. We will find finitely many cones $\mathfrak{C}(o,x_j)$, $j=1,\dots,r$, and finitely many vertices $y_1,\dots,y_s$, such that, if $v_i(x)=h_{1+\varepsilon}(x)\chi_{\mathfrak{C}(o,x_i)}$, then $h_z(g_1g)h_{1/2-it}(gg_2)$ is, after correction on a compact set, a linear combination of the functions $v_i(gy_j)$. A perfectly analogous statement will be true for $h_z(g_1g)h_{1/2+it}(gg_2)$ if the functions v_i are replaced by the functions $u_i=h_{1+\varepsilon+2it}\chi_{\mathfrak{C}(o,x_i)}$.

We may write $h_z(g_1g)h_{1/2-it}(gg_2)$

$$= q^{-(1+\varepsilon)d(go,o)}q^{-z(d(g_1go,o)-d(go,o))}$$

$$q^{-(1/2-it)(d(gg_2o,o)-d(go,o))}.$$

The function $d(g_1go,o)-d(go,o)$ is constant for $g\in\mathfrak{C}(o,x)$, as soon as $d(x,o)\geq d(g_1o,o)$. Indeed, if $d(g_1o,o)=d(g_1^{-1}o,o)\leq d(x,o)$, and $go\in\mathfrak{C}(o,x)$, then the chain connecting $g_1^{-1}o$ to go goes through x. Therefore

$$d(g_1go,o)-d(go,o) = d(go,g_1^{-1}o)-d(go,o)$$

$$= d(go,x)+ d(x,g_1^{-1}o) - d(go,x)-d(o,x) = d(x,g_1^{-1}o)- d(o,x),$$

which is independent of g. Similar reasoning shows that $d(gg_2o,o)-d(go,o)$ is constant when $g^{-1}o\in\mathfrak{C}(o,x)$, as soon as $d(x,o)>d(g_2o,o)$. Let N be the greater of the two integers $d(g_1o,o)$ and $d(g_2o,o)$ and let $x_1,\dots,x_r$ be the points at distance N from o; let $E_{ij}=\{g\colon go\in\mathfrak{C}(o,x_i)\}\cap\{g\colon g^{-1}o\in\mathfrak{C}(o,x_j)\}$. Then outside the compact set $\{g\colon d(go,o)\leq N\}$ the functions $h_z(g_1g)h_{1/2+it}(gg_2)$ and $h_z(g_1g)h_{1/2-it}(gg_2)$ are respectively linear combinations of the functions $h_{1+\varepsilon}(g)\chi_{E_{ij}}(g)$ and $h_{1+\varepsilon+2it}(g)\chi_{E_{ij}}(g)$. We shall consider one of the functions above, relative to two vertices x_1 and x_2 at distance N from o. We fix then $x'_i\in[o,x_i]$ such that $d(x_i,x'_i)=1$, for i=1,2. Let A= {g: $gx_2\in\mathfrak{E}(o,x_1)$} and B={g: $gx'_2\in\mathfrak{E}(o,x_1)$}. Then

$$h_{1+\varepsilon}(g)\chi_{E_{12}}(g)$$
$$=(1+q^{-2(1+\varepsilon)})^{-1}h_{1+\varepsilon}(x_2)(h_{1+\varepsilon}(gx_2)\chi_A - q^{-(1+\varepsilon)}h_{1+\varepsilon}(gx'_2))\chi_B).$$

A similar formula holds for $h_{1+\varepsilon+2it}\ \chi_{E_{12}}$, if $1+\varepsilon$ on the right hand side is replaced by $1+\varepsilon+2it$. But $\chi_A(g)=\chi_{\mathfrak{E}(o,x_1)}(gx_2)$ and $\chi_B(g)=\chi_{\mathfrak{E}(o,x_1)}(gx'_2)$. Therefore $h_{1+\varepsilon}\chi_{E_{12}}$ as a function of g may be written as a linear combination of functions of the type $v_i(gy_j)$ where $v_i(x)=h_{1+\varepsilon}(x)\chi_{\mathfrak{E}(o,x_i)}(x)$. We conclude that $h_z(g_1g)h_{1/2-it}(gg_2)$, after correction on a compact set, may be expressed as a linear combination of the functions $v_i(gy_j)$. The corresponding claim about $h_z(g_1g)h_{1/2+it}(gg_2)$ is also proved.

In conclusion, in order to prove that the expressions (2) and (3) are uniformly bounded as $\varepsilon\to0$, it suffices to show that, if x_1 and x_2 are arbitrary vertices $v(x)=h_{1+\varepsilon}(x)\chi_{\mathfrak{E}(o,x_1)}(x)$ and $u(x)=h_{1+\varepsilon+2it}(x)\chi_{\mathfrak{E}(o,x_1)}(x)$,

$$|G/\Gamma|\sum_{\gamma\in\Gamma} v(\gamma x_2)-\int_G v(gx_2)\ dg=O(1), \qquad (4)$$

$$|G/\Gamma|\sum_{\gamma\in\Gamma} u(\gamma x_2)-\int_G u(gx_2)\ dg=O(1). \qquad (5)$$

Observe that

$$\int_G v(gx_2)\ dg = \sum_{x\in\mathfrak{X}} v(x)m(\{g: gx_2=x\})$$

where m is the Haar measure of G. Notice that, if $g_1o=x$ and $g_2o=x_2$, $\{g: gx_2=x\}=\{g: g_1^{-1}gg_2o=o\}=g_1Kg_2^{-1}$. Therefore m({g: gx_2= x })=1. It follows that $\int_G v(gx_2)dg = \sum_{x\in\mathfrak{X}} v(x)$. We must prove therefore that

$$|G/\Gamma| \sum_{\gamma\in\Gamma} v(\gamma x_2) - \sum_{x\in\mathfrak{X}} v(x) = O(1)$$

and

$$|G/\Gamma| \sum_{\gamma\in\Gamma} u(\gamma x_2) - \sum_{x\in\mathfrak{X}} u(x) = O(1)$$

(9.1) *LEMMA. Let* v *and* u *be the functions defined above. Let* $\mu=\mu(1+2it)=(q+1)^{-1}(q^{1+2it}+q^{-2it})$. *Let* L *be the Laplace operator on* $\mathfrak{X}$. *Then*

$$\sum_{x\in\mathfrak{X}} |v(x) - Lv(x)| = O(1), \qquad (6)$$

$$\sum_{x\in\mathfrak{X}} |\mu u(x) - Lu(x)| = O(1). \qquad (7)$$

PROOF. If $x\notin\mathfrak{C}(\mathbf{o},x_1)$ and $\mathbf{d}(x,x_1)>1$, then $v(x)-Lv(x)=u(x)-Lu(x)=0$. Since there are only a finite number of vertices at distance not greater than 1, it suffices to show that

$$\sum_{\substack{d(x,x_1)>1 \\ x\in\mathfrak{C}(o,x_1)}} |v(x)-Lv(x)| = O(1), \text{ and } \sum_{\substack{d(x,x_1)>1 \\ x\in\mathfrak{C}(o,x_1)}} |\mu u(x)-Lu(x)| = O(1).$$

But, if $\mathbf{d}(x,x_1)>1$ and $x\in\mathfrak{C}(\mathbf{o},x_1)$, $v(x)=q^{-(1+\varepsilon)d(\mathbf{o},x)}=h_{1+\varepsilon}(x)$ and $Lv(x)=Lh_{1+\varepsilon}(x)=(q+1)^{-1}(q^{1+\varepsilon}+q^{-\varepsilon})h_{1+\varepsilon}(x)$. Therefore $v(x)-Lv(x)= h_{1+\varepsilon}(x)(1-(q+1)^{-1}(q^{1+\varepsilon}+q^{-\varepsilon})$ and

$$\sum_{\substack{d(x,x_1)>1 \\ x\in\mathfrak{C}(o,x_1)}} |v(x)-Lv(x)| =$$

$$|1-(q+1)^{-1}(q^{1+\varepsilon}+q^{-\varepsilon})|h_{1+\varepsilon}(x_1) \sum_{n=2}^{\infty} q^n q^{-(1+\varepsilon)n}$$

$$= (q+1)^{-1}(1-q^{-\varepsilon})^{-1}q^{-2\varepsilon}(q(1-q^{\varepsilon})+(1-q^{-\varepsilon}))\, h_{1+\varepsilon}(x_1).$$

As $\varepsilon\to 0$, the expression above converges to $(q-1)(q+1)^{-1}h_1(x_1)$. This proves the estimate (6). As for (7), observe that, for $u\in\mathfrak{C}(\mathbf{o},x_1)$ and $\mathbf{d}(x,x_1)>1$, $Lu(x)=(q+1)^{-1}(q^{1+\varepsilon+2it}+q^{-\varepsilon-2it})u(x) =\mu(1+2it+\varepsilon)u(x)$. Therefore,

$$|\mu u(x)-Lu(x)|=|\mu(1+2it+\varepsilon) - \mu(1+2it)|\,|u(x)|$$

$$= (q+1)^{-1}(q^{1+2it}(q^{\varepsilon}-1) + q^{-2it}(q^{-\varepsilon}-1))\,|u(x)| \le M\varepsilon.$$

But $\sum_{d(x,x)>1} |u(x)| = h_{1+\varepsilon}(x_1)q^{-2\varepsilon}/(1-q^{-\varepsilon})$, and the estimate (7) follows. ∎

Given a summable function on $\mathfrak{X}$ it is possible to define a function F on the set of Γ-orbits on $\mathfrak{X}$, as $F(\Gamma x) = \sum_{\gamma\in\Gamma} f(\gamma x)$. Since the function v is summable on $\mathfrak{X}$ we may define in this fashion on the set of orbits $\Gamma\backslash\mathfrak{X}$ the function $V(\Gamma x) = \sum_{\gamma\in\Gamma} v(\gamma x)$, and a similar function U defined on $\Gamma\backslash\mathfrak{X}$ associated to u. Therefore (4) and (5), keeping in mind that $|G/\Gamma|$ is exactly the number of orbits $|\Gamma\backslash\mathfrak{X}|$, may be expressed as

$$V(\Gamma x_2) - 1/|\Gamma\backslash\mathfrak{X}| \sum_{\Gamma x\in\Gamma\backslash\mathfrak{X}} V(\Gamma x) = O(1), \tag{8}$$

$$U(\Gamma x_2) - 1/|\Gamma\backslash\mathfrak{X}| \sum_{\Gamma x\in\Gamma\backslash\mathfrak{X}} U(\Gamma x) = O(1). \tag{9}$$

In other words we must show that, as $\varepsilon\to 0$, the values of the functions U and V at a given point of $\Gamma\backslash\mathfrak{X}$ are not too distant from their average values on the entire set of orbits.

To prove (8) and (9) we shall introduce a Laplace operator for functions defined on $\Gamma\backslash\mathfrak{X}$. We define

$$\tilde{L}F(\Gamma x) = (q+1)^{-1} \sum_{d(x,y)=1} F(\Gamma y).$$

Observe that $\tilde{L}$ may be thought of as the ordinary Laplace operator defined on Γ-invariant functions on $\mathfrak{X}$. A basis for the finite-dimensional vector space of complex-valued functions defined on $\Gamma\backslash\mathfrak{X}$ is given by the functions $\delta_{\Gamma x}$ which are 1 on the orbit Γx and zero on every other orbit. With respect to this basis the matrix coefficients of $\tilde{L}$ are given by

$$(\tilde{L}\delta_{\Gamma x}, \delta_{\Gamma y}) = (q+1)^{-1} \sum_{d(y,z)=1} \delta_{\Gamma x}(\Gamma z) = (q+1)^{-1} |\{\gamma:\ \mathbf{d}(\gamma x, y)=1\}|.$$

But $\mathbf{d}(\gamma x, y) = \mathbf{d}(x, \gamma^{-1}y)$, and therefore $\tilde{L}$ has a symmetric matrix with nonnegative elements. We shall now use a classical result on nonnegative matrices [S], for which we need a definition.

(9.2) *DEFINITION. A nonnegative self-adjoint matrix* A *is called irreducible if for every pair of indices* (i,j) *there exists* k *such that* $(A^k)_{ij}>0$.

We observe that $\tilde{L}$ is irreducible with respect to its canonical basis. Indeed, if Γx and Γy are given, there exists k such that $(L^k\delta_x,\delta_y)>0$ (it suffices to take $k=\mathbf{d}(x,y)$). It follows then that $(\tilde{L}^k\delta_{\Gamma x},\delta_{\Gamma y})>0$.

(9.3) *THEOREM. Let* A *be a nonnegative irreducible self-adjoint matrix; then there exists a unique eigenvalue* r *such that*

(a) r *is a real positive number,*

(b) an eigenvector of r *has strictly positive entries,*

(c) $r\geq|\lambda|$ *for every eigenvalue* λ,

(d) the eigenspace associated to r *is one-dimensional,*

(e) if λ *is any eigenvalue and* $|\lambda|=r$, *then* λ *is a simple root of the characteristic equation of* A.

PROOF. [S, Theorem 1.5 and Theorem 1.7].

We now apply the previous theorem to the matrix $\tilde{L}$. Observe that, if **1** is the function identically 1 on $\Gamma\backslash\mathfrak{X}$, then $\tilde{L}\mathbf{1}=\mathbf{1}$, because $(\tilde{L}\delta_{\Gamma x},\delta_{\Gamma y})=(q+1)^{-1}|\{\gamma\in\Gamma:\ \mathbf{d}(\gamma x,y)=1\}|$, and $\sum_{\Gamma y}(\tilde{L}\delta_{\Gamma x},\delta_{\Gamma y})$ $=(q+1)/(q+1)=1$. On the other hand the norm of $\tilde{L}$ is 1. Therefore 1 is the maximum eigenvalue of $\tilde{L}$. Since the eigenspace of **1** has dimension 1, the operator $(I-\tilde{L})^{-1}$ is bounded on the subspace of $\ell^2(\Gamma\backslash\mathfrak{X})$ which is orthogonal to **1**, that is on the space of functions which have mean value zero. Denote this space by $\mathfrak{M}$. We can then write the function V of (8) as the sum of two orthogonal functions V_1 and V_2, where V_1 is the constant function $|\Gamma\backslash\mathfrak{X}|^{-1}\sum_{\Gamma x} V(\Gamma x)$, and V_2 has mean value zero. Since $(I-L)$ is invertible on $\mathfrak{M}$, and because of (9.1(6)), $\|V_2\|\le C\|(I-\tilde{L})V_2\|=C\|(I-\tilde{L})V\|=O(1)$. Therefore $\|V-V_1\|=\|V_2\|=O(1)$. The space $\ell^2(\Gamma\backslash\mathfrak{X})$ is finite-dimensional, and therefore this implies that, for every x_2, $V(\Gamma x_2)-V_1=O(1)$, in other words (8) holds. In order to show that (9) is true, we observe that, if $\mu=\mu(1+2it)=(q^{1+2it}+q^{-2it})/(q+1)$, then under the hypothesis of the theorem μ is a strictly complex number. Therefore $(\mu-\tilde{L})$ is invertible. Therefore (9.1(7)) implies that $\|U\|\le C\|(\mu-\tilde{L})U\|=O(1)$. This implies that both terms on the left-hand side of (9) are bounded and therefore the estimate holds. We have thus concluded the proof of STEP B.

10. The constant 1 is a cyclic vector (STEP C). In order to complete the proof of (7.2) we must now show that **1** is a cyclic vector in $L^2(\Omega,\nu)$ for the representation π_Γ. In other words we must prove the following.

(10.1) *PROPOSITION. Let ξ be a nonzero element of $L^2(\Omega,\nu)$, then there exists at least one element $\gamma\in\Gamma$, such that*

$$(\xi,\pi(\gamma)\mathbf{1})=\int_\Omega P^{1/2-it}(\gamma,\omega)\xi(\omega)\,d\nu \neq 0.$$

Recall that the Poisson transform relative to the complex number $z=1/2-it$ is defined on the space of finitely additive measures as $\mathcal{P}_z(m)(x)=\int_\Omega P^{1/2-it}(x,\omega)dm$. Given an element ξ in $L^2(\Omega,\nu)$, we may apply $\mathcal{P}_z$ to the measure $\xi d\nu$, and define, for $x\in\mathfrak{X}$,

$$\mathcal{P}\xi(x) = \mathcal{P}_z(\xi d\nu)(x) = \int_\Omega P^{1/2-it}(x,\omega)\xi(\omega)\,d\nu.$$

In order to prove (10.1) we must show that, if $\mathcal{P}\xi(\gamma o)=0$ for every $\gamma\in\Gamma$, then $\xi=0$.

Recall also the definition of the *intertwining operator*

I_z, introduced in Section 3, for $z\in\mathbb{C}$. For simplicity of notation we set $I_{1/2+it}=I_t$ and recall that I_t is a unitary operator on $L^2(\Omega,\nu)$, with the property that

$$\int_\Omega P^{1/2-it}(x,\omega)\xi(\omega)\,d\nu = \int_\Omega P^{1/2+it}(x,\omega)I_t\xi(\omega)\,d\nu.$$

We now apply (1.3) to obtain a convenient expression for $\mathcal{P}\xi$. Let $x\neq 0$, and, as in the statement of (1.3), let x' be the vertex at distance 1 from x in the chain $[o,x]$; then an application of (1.3) yields

$$\mathcal{P}\xi(x) - q^{-(1/2+it)}\mathcal{P}\xi(x') =$$

$$(q^{1/2+it}-q^{-1/2-it})q^{(1/2+it)d(x',o)}\int_{\Omega(x)}\xi(\omega)d\nu.$$

Likewise,

$$\mathcal{P}\xi(x) - q^{-(1/2-it)}\mathcal{P}\xi(x') =$$

$$(q^{1/2-it}-q^{-1/2+it})q^{(1/2-it)d(x',o)}\int_{\Omega(x)}I_t\xi(\omega)d\nu.$$

Multiplying the first equation by $q^{1/2+it}$, and the second by $q^{(1/2-it)}$, and subtracting, one obtains, $\mathcal{P}\xi(x)=$

$$\frac{(q^{1/2-it}-q^{-1/2+it})}{(q^{-1/2-it}-q^{-1/2+it})}(q+1)^{-1}\Bigl[q^{-(1/2+it)d(o,x)}\nu(\Omega(x))^{-1}\int_{\Omega(x)}\xi(\omega)d\nu$$

$$- q^{-(1/2-it)d(o,x)}\nu(\Omega(x))^{-1}\int_{\Omega(x)}I_t\xi(\omega)d\nu\Bigr].$$

In other words,

$$\mathcal{P}\xi(x) = C(t)\Bigl[q^{-(1/2+it)d(o,x)}\nu(\Omega(x))^{-1}\int_{\Omega(x)}\xi(\omega)d\nu -$$

$$q^{-(1/2-it)d(o,x)}\nu(\Omega(x))^{-1}\int_{\Omega(x)}I_t\xi(\omega)d\nu\Bigr]. \tag{1}$$

(10.2) *LEMMA. Let ξ be a square-integrable function on Ω; then, for almost every $\omega\in\Omega$,*

$$\lim_{x\to\omega} \nu(\Omega(x))^{-1}\int_{\Omega(x)} \xi(\omega)d\nu = \xi(\omega),$$

as x approaches ω keeping within a bounded distance from $[\mathbf{o},\omega)$ (nontangential convergence).

PROOF. Let $[\mathbf{o},\omega)$ be the chain $\{\mathbf{o},x_1,x_2,\dots\}$. Then the Vitali-Lebesgue theorem **[R]** directly implies that, for almost every ω,

$$\lim \nu(\Omega(x_n))^{-1}\int_{\Omega(x_n)} \xi(\omega)\, d\nu = \xi(\omega)\ .$$

Let $\mathbf{d}(x,[\mathbf{o},\omega))=\mathbf{d}(x,x_n)=m<M$, where M is a fixed positive integer; then, for $\omega\in\Omega(x)\subseteq\Omega(x_n)$, and, with the notation introduced at the end of Section 5,

$$|\nu(\Omega(x))^{-1}\int_{\Omega(x)} \xi(\omega)\, d\nu - \nu(\Omega(x_n))^{-1}\int_{\Omega(x_n)} \xi(\omega)\, d\nu|$$

$$= |\Delta_{m+n}\xi(\omega)+\Delta_{m+n-1}\xi(\omega)+\dots+\Delta_n\xi(\omega)| \le (m+1) \sup_{0\le k\le m} |\Delta_{n+k}\xi(\omega)|$$

$$\le M\sqrt{q} \sup_{0\le k\le m} \|\Delta_{n+k}\xi\| \to 0.$$

This concludes the proof of the lemma. ∎

It follows now that (1) implies the estimate

$$q^{\mathbf{d}(\mathbf{o},x)/2}\mathcal{P}\xi(x)=(q^{-it\mathbf{d}(\mathbf{o},x)}\xi(\omega)-\ q^{+it\mathbf{d}(\mathbf{o},x)}I_t\xi(\omega))+o(1) \qquad (2)$$

for almost every $\omega\in\Omega$, when x approaches ω nontangentially.

Choose now $\omega_o\in\Omega$, such that $\xi(\omega_o)\neq 0$. Keeping in mind the estimate (2), in order to prove (10.1) we must show that, as $\mathbf{d}(\gamma\mathbf{o},\mathbf{o})$ becomes large, subject, for some M, to the condition $\mathbf{d}(\gamma\mathbf{o},[\mathbf{o},\omega_o))\le M$, the quantity

$$q^{-it\mathbf{d}(\mathbf{o},\gamma\mathbf{o})}\xi(\omega_o)-q^{+it\mathbf{d}(\mathbf{o},\gamma\mathbf{o})}I_t\xi(\omega_o) \qquad (3)$$

does not become arbitrarily small. We may assume that $|\xi(\omega_o)|=|I_t\xi(\omega_o)|$. Indeed, if not, (3) is bounded away from zero by $||\xi(\omega_o)|-|I_t\xi(\omega_o)||$, and cannot become arbitrarily small. Let $q^{it}=e^{i2\pi\theta}$. Then the quantity (3) may be written as a constant multiple of $\sin(2\pi\theta\mathbf{d}(\gamma\mathbf{o},\mathbf{o})+\theta_0)$, where $\theta_0=(\theta_1-\theta_2)/2$, and θ_1 and θ_2 are respectively the arguments of $\xi(\omega_o)$ and $I_t\xi(\omega_o)$. In other words we are reduced to showing that there exist $\varepsilon>0$, and a positive integer M, such that, for infinitely many γ satisfying $\mathbf{d}(\gamma\mathbf{o},[\mathbf{o},\omega_o))\le M$,

$$|\sin(2\pi\theta\mathbf{d}(\gamma\mathbf{o},\mathbf{o})+\theta_0)|\ge \varepsilon. \qquad (4)$$

We will consider now two different cases.

Case (1): the number θ is irrational. In this case we first choose a number M' large enough that infinitely many points of $\Gamma\mathbf{o}$ satisfy $\mathbf{d}(\gamma\mathbf{o},[\mathbf{o},\omega))\le M'$. This is possible because $\Gamma\backslash\mathfrak{X}$ is finite, and a fundamental region of Γ (a set containing exactly one element of each orbit of Γ) is likewise finite. We let $S=\Gamma\cap\{x\colon \mathbf{d}(x,[\mathbf{o},\omega))\le M'\}$. Next, again using the fact that Γ has finitely many orbits, we choose N so that each of the q+1 disjoint cones $\mathfrak{C}(\mathbf{o},x_j)$, where $x_1,\dots,x_{q+1}$ are the nearest neighbors of $\mathbf{o}$, contains an element of $\Gamma\mathbf{o}$ at distance no greater than N from $\mathbf{o}$. Finally we let M=M'+N. By construction every ball of radius N centered at an element $\gamma\mathbf{o}$ of S contains elements of $\Gamma\mathbf{o}$ inside each cone $\mathfrak{C}(\gamma\mathbf{o},\gamma x_j)$. This means that for each $\gamma\mathbf{o}\in S$ there is a vertex $\gamma'\mathbf{o}$, such that the chain $[\mathbf{o},\gamma\mathbf{o}]$ is properly contained in $[\mathbf{o},\gamma'\mathbf{o}]$, and $\mathbf{d}(\gamma\mathbf{o},\gamma'\mathbf{o})\le N$. Now, for $\alpha\in\mathbb{R}$, define $f_n(\alpha)=\sin(\alpha+2\pi n\theta)$. The fact that θ is irrational implies that $f_n(0)\ne 0$. Let δ be such that, for $n=1,\dots,N$ $|f_n(0)|>\delta>0$. Choose $\varepsilon>0$, such that $\delta>\varepsilon$, and $|f_n(\alpha)|>\delta>\varepsilon$, for $n=1,\dots,N$ whenever $|\sin\alpha|<\varepsilon$. Let $\gamma\mathbf{o}$ be an element of S such that (4) fails and let $\alpha=2\pi\theta\mathbf{d}(\gamma\mathbf{o},\mathbf{o})+\theta_0$. Then $|\sin\alpha|<\varepsilon$, and therefore, for every $n=1,\dots,N$ $|\sin(\alpha+2\pi n\theta)|>\varepsilon$. But by (10.3) there exists $\gamma'\mathbf{o}$, such that $\mathbf{d}(\gamma'\mathbf{o},[\mathbf{o},\omega))\le M$, and $\mathbf{d}(\gamma'\mathbf{o},\mathbf{o})=\mathbf{d}(\gamma\mathbf{o},\mathbf{o})+n$, with $0<n\le N$. It follows that $2\pi\theta\mathbf{d}(\gamma'\mathbf{o},\mathbf{o})+\theta_0=\alpha+2\pi n\theta$, and therefore $|\sin(2\pi\theta\mathbf{d}(\gamma'\mathbf{o},\mathbf{o})+\theta_0)|\ge\varepsilon$. We have thus proved that for every element $\gamma\mathbf{o}\in S$ which fails to satisfy (4) there exists at least one other element $\gamma'\mathbf{o}\in\Gamma\mathbf{o}$, still within distance M from the

infinite chain $[o,\omega)$, and at distance N from γo, which satisfies (4). This concludes the proof for the case in which θ is irrational.

Case (2): the number θ *is rational,* $\theta=r/s$. We may assume that r and s are integers with no common divisors. The hypothesis $q^{2it}\neq\pm1$ of (7.2) implies that $s\neq2$. A careful scrutiny of the proof for the case in which θ is rational shows that the same arguments work if we can prove that each of the $q+1$ cones $\mathfrak{C}(o,x_j)$ contains vertices $\gamma_j o$, such that $d(o,\gamma_j o)$ is not an integral multiple of s. We shall presently prove that this is indeed the case. We shall treat a slightly more general case.

(10.3) *DEFINITION. A subgroup of* $\mathrm{Aut}(\mathfrak{X})$ *is called cofinal if, for every* $x\in\mathfrak{X}$, *the set* $\mathfrak{C}(o,x)\cap\Gamma o$ *is nonempty.*

Observe that the hypothesis that Γ is cocompact implies that it is cofinal, because a cone has infinite diameter.

(10.4) *LEMMA. Let* $\Gamma\leq\mathrm{Aut}(\mathfrak{X})$ *be a cofinal sugroup. Suppose that there exist an integer* $s\geq3$, *and an element* $x_o\in\mathfrak{X}$, *such that* $d(\gamma o,o)$ *is an integral multiple of* s *for every* $\gamma o\in\mathfrak{C}(o,x_o)$. *Then there exists* $\omega_o\in\Omega$, *such that* $\Gamma\omega_o=\{\omega_o\}$.

PROOF. Let $\mathfrak{U}=\{\gamma^{-1}o\colon \gamma\in\Gamma,\ \gamma o\in\mathfrak{C}(o,x_o)\}$. We shall prove first of all that this set has exactly one limit point $\omega_o\in\Omega$. It suffices to show that given $\gamma_1,\gamma_2\in\mathfrak{U}$, such that $d(\gamma_j o,o)\geq d(x_o,o)+N+1$, for $j=1,2$, $[o,\gamma_1^{-1}o]\cap[o,\gamma_2^{-1}o]$ is a chain of at least N edges. As N becomes large this identifies an infinite chain $[o,\omega_o)$. To prove the assertion suppose that $[o,\gamma_1^{-1}o]\cap[o,\gamma_2^{-1}o]$ consists of N_o edges with $N_o<N$. Find a vertex x such that $d(o,x)=N_o+2$, and such that the chain $[o,x]$ has exactly N_o edges in common with $[o,\gamma_1^{-1}o]$ and N_o+1 edges in common with $[o,\gamma_2^{-1}o]$. Let γ be such that $\gamma o\in\mathfrak{C}(o,x)$ (Fig. 2).

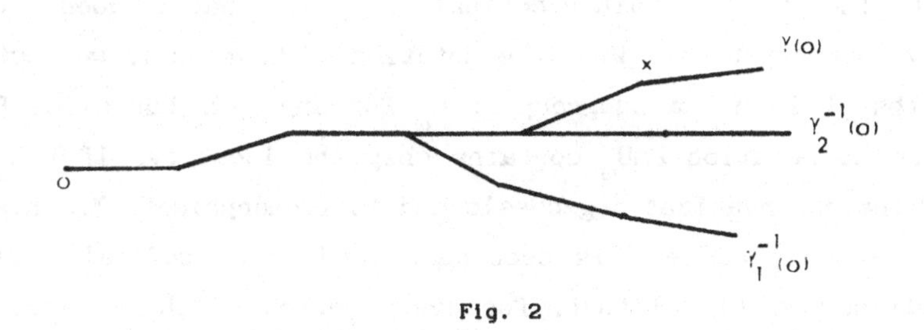

Fig. 2

Observe (Fig. 2) that $d(\gamma_1\gamma o,o)=d(\gamma o,\gamma_1^{-1}o)=d(o,\gamma o)+d(o,\gamma_1^{-1}o)$ $-2N_o$, by construction of x and γ. This means that $d(\gamma_1\gamma o,o)=$ $d(\gamma o,o)-2N_o$ (mod s). A similar argument leads to the conclusion that $d(\gamma_2\gamma o,o)=d(\gamma o,o)-2(N_o+1)$ (mod s). We conclude that $d(\gamma_1\gamma o,o)\neq d(\gamma_2\gamma o,o)$ (mod s), and therefore $\gamma_1\gamma o$ and $\gamma_2\gamma o$ cannot both belong to $\mathfrak{C}(o,x_o)$. We will reach a contradiction and prove the assertion if we show, on the contrary, that $\gamma_1\gamma o$ and $\gamma_2\gamma o$ are in $\mathfrak{C}(o,x_o)$. Apply now γ_j to the chains $[o,\gamma_j^{-1}o]$ and $[\gamma_j^{-1}o,\gamma o]$ (Fig. 2). We obtain the chains $[o,\gamma_j o]$ and $[o,\gamma_j\gamma o]$ (Fig. 3).

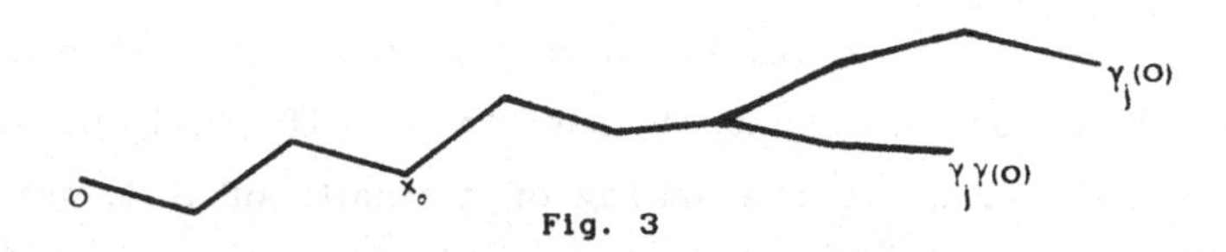

Fig. 3

The total number of edges of $[o,\gamma_j o]$ is at least $d(x_o,o)+N+1$, while only N_o edges lie in the portion of the chain $[o,\gamma_j o]$ which is not in common with $[o,\gamma_j\gamma o]$ (Fig. 3). It follows that x_o must be in the portion of the chain $[o,\gamma o]$ which is in common with $[o,\gamma_j\gamma o]$. In other words, $\gamma_j\gamma o\in\mathfrak{C}(o,x_o)$. This contradiction proves our assertion. We let ω_o be the limit point of $\mathfrak{U}$. If $\gamma'\in\Gamma$, then $\gamma'\omega_o=\lim\{\gamma'\gamma^{-1}o\colon \gamma\in\Gamma,\ \gamma o\in\mathfrak{C}(o,x_o)\}=$ $\lim\{\gamma^{-1}o\colon \gamma\gamma' o\in\mathfrak{C}(o,x_o)\}$. But $\gamma\gamma' o\in\mathfrak{C}(o,x_o)$, when $\gamma o\in\mathfrak{C}(o,x_o)$, as soon as $d(\gamma o,o)\geq d(o,x_o)+d(\gamma' o,o)$. It follows that $\gamma'\omega_o=\omega_o$. ∎

We go back now to the hypothesis of (7.2) for the discrete group Γ. Then Γ is certainly cofinal. We prove that it does not fix any boundary point. With the notation of Chapter I, we must prove that Γ is not a subgroup of G_ω for any $\omega\in\Omega$. But since Γ contains no rotation $\Gamma\cap B_\omega$ contains only the identity. If $\Gamma\subseteq G_\omega$ then Γ leaves invariant a geodesic and is isomorphic to $\mathbb{Z}$. This cannot happen because Γ is cocompact (and hence cofinal). We conclude now by (10.4) that, for every vertex x, there exists $\gamma\in\Gamma$, such that $\gamma\mathbf{o}\in\mathfrak{C}(\mathbf{o},x)$ and $\mathbf{d}(\gamma\mathbf{o},\mathbf{o})\neq 0$ (mod s). This proves (10.1) and concludes STEP C, and the proof of (7.2).

11. Notes and remarks. Theorem (1.2) is essentially the same as Theorem A of **[MZ1]** dealing with eigenfunctions of the simple random walk on a free group. The simpler proof given here which is based only on the geometry of the tree is taken from **[F-T S2]** where a different and more general result is proven for a nearest-neighbor anisotropic random walk on a homogeneous tree. The results of **[K1,2]** on affine buildings of arbitrary rank are also more general but in a different direction. Observe that (1.2) is the analog of a result of S. Helgason for symmetric spaces **[He]**. Proposition (1.3) is a special case of Lemma 2.1 of **[F-T S2]**. Related to this proposition are the more general results of **[KP]**, **[KPT]** and **[PW]**. The theory of spherical functions and spherical representations as described in Section 2 and 4 is classical. We have especially benefited from the expositions given in **[L]** and **[F]**. Spherical functions and spherical representations for a group acting transitively on a tree and its boundary were first considered by **[C2]**, but spherical functions for **PGL**(2) over a p-adic field were studied by G. van Dijk **[Di]**. Observe that spherical representations and spherical functions in this context are nothing but the restrictions of spherical representations and spherical functions of the full group Aut($\mathfrak{X}$). Indeed much of the theory may be developed intrinsically for the free group or any other group which acts simply transitively on the vertices of the tree [**F-T P1,2**]. An interesting characterization of spherical representations of free groups in terms of a condition of symmetry of a coefficient viewed as a function on $\mathfrak{X}$ is given in [**KS2**]. The explicit formulas contained in (2.3) and (2.4) are taken from [**F-T P1,2**]. Intertwining operators were used explicitly by [**MZ1**] in order to define the representations of the complementary series. However the convenient diagonalization of the intertwining operators given in Lemma (3.2) is taken from [**F-T S2**] where this result is proved in a more general context. The inequalites of Lemma (6.1) and Lemma (6.2) are particular instances of more general convolution theorems due to U. Haagerup [**H**]. The formula of Proposition (6.3) for the explicit computation of the resolvent was given in [**F-T P1,2**]. The spherical Plancherel formula was first computed by Cartier [**C3**]. A different computation in the context of free groups was given in [**F-T P1,2**] (see also [**CdM**]). Our immediate source for (6.4) is the simple computation given in [**FP**]. The direct ancestor of Theorem (7.1) is the theorem on the irreducibility of spherical representations of free groups [**F-T P1,2**]. One should mention however the earlier result of T. Pytlik [**P**] yielding irreducibility for almost every representation of the principal series of a free group. Related results are the theorem on the irreducibility of the representations associated to anisotropic nearest-neighbor random walks [**F-T S2**], and the theorem of finite reducibility for representations associated to an arbitrary group-invariant symmetric finitely supported random walk [**St1**]. An exposition of the latter result is contained in [**A**]. Theorem (7.1) has an analog for lattices of SL(2,$\mathbb{R}$) [**St3**]. An important rigidity theorem which makes use of

(7.1) is the following.

Let $\Gamma_1=i_1(\Gamma)$ *and* $\Gamma_2=i_2(\Gamma)$ *be two embeddings of a free group* Γ *as cocompact subgroups of* $\mathrm{Aut}(\mathfrak{X})$. *Let* π_1 *and* π_2 *be the restrictions to* Γ_1 *and* Γ_2, *respectively, of a spherical representation of* $\mathrm{Aut}(\mathfrak{X})$. *Suppose that* π_1 *is unitarily equivalent to* π_2. *Then for some* $g\in\mathrm{Aut}(\mathfrak{X})$ $g\Gamma_1g^{-1}=\Gamma_2$.

This follows from the techniques of **[BS1,2]** (which contains the analogous result for embeddings of a free group in $SL(2,\mathbb{R})$) and Theorem 3.7 of **[CM]**.

Observe that this implies that spherical representations of a free group, as defined in **[F-T P1,2]**, depend on the choice of the generators. In other words if two sets of generators are not mapped one into the other by an inner automorphism of the free group they give rise to inequivalent spherical representations.

We now describe briefly the tools required for the proof of (7.1) in addition to those which were used for the proof of (7.2).

The hypothesis that $\{\gamma\in\Gamma\colon \gamma x=x\}=\{e\}$ is inessential because of a result of H. Bass and R. Kulkarni **[BKu]** which asserts that every discrete cocompact subgroup of $\mathrm{Aut}(\mathfrak{X})$ contains a subgroup of finite index satisfying the above hypothesis. Observe however that the proof does not work if we only assume that $\mathrm{Aut}(\mathfrak{X})/\Gamma$ has finite volume. Discrete groups Γ such that $\mathrm{Aut}(\mathfrak{X})/\Gamma$ has finite volume but is not compact do exist **[Lu]**. To extend the proof to representations of the complementary series we only need an estimate, analogous to (8.1) for $(\pi_s(L)-\mu)^{-1}$, when μ is a complex number and $1/2\le Rez\le 1$. This estimate holds once the spectrum of $\pi_s(L)$ is determined to be contained in an interval with end points $\pm(q^s+q^{1-s})/(q+1)$. For STEP B the hypothesis that $|\Gamma_x|=1$ is used, but not in an essential way, in some of the formulas. The hypothesis that $q^{2it}\neq\pm1$ is also used at the end of the section but it may be done away with by inverting $I+\tilde{L}$ on the complement of the eigenspace of -1. The proof of irreducibility breaks down in general at STEP C without the hypothesis that $q^{2it} \neq \pm 1$. A separate argument is needed to treat this case so as to yield (2) and (3) of (7.1).

The following general result on restrictions of irreducible representations was recently proved by M.G. Cowling and T. Steger [CS].

THEOREM. Let G *be a separable unimodular locally compact group and* Γ *be a lattice of* G *such that the quotient space* G/Γ *has finite volume. Let* ρ *be the restriction of the quasi-regular representation of* G *on* $L^2(G/\Gamma)$ *to the orthogonal complement of the space of constant functions on* G/Γ. *Let* π *be an irreducible unitary representation of* G. *If* $\pi \otimes \rho$ *does not contain* π *then the restriction of* π *to* Γ *is irreducible.*

This theorem is applied in [CS] to prove the analog of Theorem 7.1 for lattices of semisimple Lie groups. Using the same result and the theory of representations of $\mathrm{Aut}(\mathfrak{X})$ as described in Chapter III, it is also possible to establish the conclusion of Theorem (7.1) for a lattice of $\mathrm{Aut}(\mathfrak{X})$ (without the hypothesis of cocompactness). The direct proof of Theorem (7.1) given here has the advantage of not requiring any knowledge of the representation theory of $\mathrm{Aut}(\mathfrak{X})$. This suggests that its arguments may also be applied to nonspherical irreducible unitary representations of free groups and their restrictions to subgroups of finite index.

CHAPTER III

1. A classification of unitary representations. In this section we classify the irreducible unitary representations of a closed group G acting transitively on a tree and its boundary. The representations are divided into three disjoint classes: spherical representations, special representations and cuspidal representations. While all definitions make sense in this general context the class of *cuspidal* representations becomes significant only when G contains sufficiently many rotations. Accordingly, in Section 3, when cuspidal representations are studied in detail, we restrict our attention to the case $G=\mathrm{Aut}(\mathfrak{X})$. Assume now that G is a closed subgroup of $\mathrm{Aut}(\mathfrak{X})$ acting transitively on $\mathfrak{X}$ and Ω. This means by (I, 10.1) that the compact group $K_x \cap G=\{g\in G: g(x)=x\}$ acts transitively on Ω. We also notice that by [**F**, Prop. I, 1] G is unimodular because $(G, G\cap K_x)$ is a Gelfand pair (III, 4.1).

Let π be a unitary representation of G on the Hilbert space $\mathcal{H}_\pi$. We only consider representations π which are continuous in the weak operator topology, which means that, for every $\xi,\eta \in \mathcal{H}_\pi$, the function $(\pi(g)\xi,\eta)$ is continuous on G.

If K is a compact subgroup of G and dk its normalised Haar measure, we define, for $\xi \in \mathcal{H}_\pi$,

$$P_\pi(K)\xi = \int_K \pi(k)\xi \, dk .$$

It is easy to verify that $P_\pi(K)$ is an orthogonal projection and that, for each $k\in K$, $\pi(k)P_\pi(K)=P_\pi(K)$. Conversely, if $\pi(k)\xi=\xi$ for every $k\in K$, then $P_\pi(K)\xi=\xi$. In other words the range of $P_\pi(K)$ is the space of all $\pi(K)$-invariant vectors. (We shall usually refer to this space as the space of all K-invariant vectors, when the representation π is fixed.)

Let $\mathfrak{J}$ be a finite subtree of $\mathfrak{X}$. Define $K(\mathfrak{J})=\{g\in G: gx=x,$ for all $x\in\mathfrak{J}\}$ and let $P_\pi(\mathfrak{J})$ be the projection on the space of $K(\mathfrak{J})$-invariant vectors. Since $K(\mathfrak{J})$ is open in G, its Haar measure is a multiple of the Haar measure m of G restricted to $K(\mathfrak{J})$. Therefore,

$$P_\pi(\mathfrak{J}) = 1/m(K(\mathfrak{J}))\int_{K(\mathfrak{J})} \pi(g)\, dg.$$

A finite subtree $\mathfrak{J}$ is called *complete* if it consists of a single vertex, or if every vertex has degree either q+1 or 1. We denote by $\mathfrak{J}^{o}$ (the interior of $\mathfrak{J}$) the subtree of $\mathfrak{J}$ consisting of vertices of homogeneity q+1. The boundary of $\mathfrak{J}$ is the set $\partial\mathfrak{J}=\mathfrak{J}\setminus\mathfrak{J}^{o}$. Clearly every finite subtree is contained in a finite complete subtree. More precisely, for every $m>0$, $\mathfrak{J}\subset V_m(\mathfrak{J})=\{x\in\mathfrak{X}: d(x,\mathfrak{J})\le m\}$ and $V_m(\mathfrak{J})$ is a finite complete subtree. As $\mathfrak{J}$ varies among all finite complete subtrees, $K(\mathfrak{J})$ describes a basis of neighboroods of the identity of G.

It follows that, if π is a unitary representation and $0\ne\xi\in\mathcal{H}_\pi$ then, for some finite complete subtree $\mathfrak{J}$,

$$\int_{K(\mathfrak{J})} (\pi(g)\xi,\xi)\, dg \ne 0,$$

which implies $P_\pi(\mathfrak{J})\xi\ne0$. The remarks above justify the following definition.

(1.1) *DEFINITION. Let π be an irreducible unitary representation of G. Let ℓ_π be the smallest positive integer for which there exists a finite complete subtree $\mathfrak{J}$ having ℓ_π vertices and such that $P_\pi(\mathfrak{J})\ne0$. Then π is called spherical if $\ell_\pi=1$, is called special if $\ell_\pi=2$, and is called cuspidal if $\ell_\pi>2$.*

We observe that by (II,5) every spherical representation in the sense of the definition above is spherical as defined in (II,5.1). We shall therefore not discuss spherical representations in this chapter.

If $P_\pi(\mathfrak{z})\neq 0$ and $\mathrm{card}(\mathfrak{z})=\ell_\pi$ then we say that $\mathfrak{z}$ is a *minimal tree* associated to π. If $\mathfrak{z}$ is a minimal tree then $g\mathfrak{z}$ is also a minimal tree for every $g\in G$ because $P_\pi(g\mathfrak{z})=\pi(g)P_\pi(\mathfrak{z})\pi(g)^{-1}$ and $\mathrm{card}(\mathfrak{z})=\mathrm{card}(g\mathfrak{z})$.

If $\mathfrak{z}$ is any finite complete subtree, let $\mathcal{H}_\pi(\mathfrak{z})$ be the subspace of $\mathcal{H}_\pi$ consisting of all $K(\mathfrak{z})$-invariant vectors. In other words $\mathcal{H}_\pi(\mathfrak{z})$ is the range of $P_\pi(\mathfrak{z})$.

We consider now the subspace $V_\pi=\bigcup_{\mathfrak{z}}\mathcal{H}_\pi(\mathfrak{z})$. This is a nontrivial linear space. Observe that V_π is invariant under the action of $\pi(G)$ because $\pi(g)\mathcal{H}_\pi(\mathfrak{z})=\mathcal{H}_\pi(g\mathfrak{z})$ and of course $K(g\mathfrak{z})=gK(\mathfrak{z})g^{-1}$. It follows that V_π is dense in $\mathcal{H}_\pi$.

Note that the space V_π is dense for an arbitrary unitary representation π, without the hypothesis of irreducibility. Indeed, if M is a nontrivial invariant closed subspace of $\mathcal{H}_\pi$ and $\xi\neq 0$ $\xi\in\mathcal{H}_\pi$ then, for some subtree $\mathfrak{z}$, $P_\pi(\mathfrak{z})\xi\neq 0$. Thus $M\cap V_\pi\neq 0$. Since the orthogonal complement of V_π is invariant, this shows that V_π is dense.

Let $\mathfrak{z}_n$ be a sequence of finite complete subtrees such that $\mathfrak{z}_n\subset\mathfrak{z}_{n+1}$ for every n and $\mathfrak{X}=\bigcup_n\mathfrak{z}_n$. Then $\bigcup_n\mathcal{H}_\pi(\mathfrak{z}_n)=V_\pi$. If, in particular, $\dim(\pi)<+\infty$, then V_π is closed and $\mathcal{H}_\pi=\mathcal{H}_\pi(\mathfrak{z}_{n_o})$ for some n_o; this means that $\pi|_{K(\mathfrak{z}_{n_o})}$ is trivial. We have thus proved the following proposition.

(1.2) *PROPOSITION. Let π be a unitary continuous representation of* G. *If* $\dim(\pi)<+\infty$ *then there exists a complete finite subtree* $\mathfrak{z}$ *such that* $\pi|_{K(\mathfrak{z})}$ *is trivial.*

Let π be a unitary representation and let $\mathfrak{z}$ be a finite complete subtree such that the number of vertices of $\mathfrak{z}$ is exactly ℓ_π. This means that if $\mathfrak{y}$ is a proper complete subtree of $\mathfrak{z}$, then $P_\pi(\eta)\xi=0$ for every $\xi\in\mathcal{H}_\pi(\mathfrak{z})$. It follows that, if $\xi,\eta\in V_\pi$ and $P_\pi(\mathfrak{z})\xi=\xi$, the coefficient function $u(g)=(\pi(g)\xi,\eta,)$ has the following properties:

(1) $u(gk) = (\pi(g)\pi(k)\xi,\eta) = u(g)$, *for* $k\in K(\mathfrak{z})$;

(2) *there exists* $\mathfrak{z}'$ *such that* $u(kg) = (\pi(g)\xi,\pi(k^{-1})\eta) = u(g)$ *for all* $k\in K(\mathfrak{z}')$ *($\mathfrak{z}'$ is any complete subtree for which $P_\pi(\mathfrak{z}')\eta=\eta$);*

(3) *if* $\mathfrak{s}$ *is a complete subtree properly contained in* $\mathfrak{z}$ *then*

$$1/m(K(\mathfrak{s}))\int_{K(\mathfrak{s})} u(gk)\,dk = (P_\pi(\mathfrak{s})\xi,\pi(g^{-1})\eta)=0 \text{ for all } g\in G.$$

The remarks above justify the following definition.

(1.3) *DEFINITION. Let* $\mathfrak{z}$ *be a complete finite subtree of* $\mathfrak{X}$. *We let* $\mathcal{S}(\mathfrak{z})$ *be the linear space of continuous functions* u *satisfying the following properties:*

(1) u *is* $K(\mathfrak{z})$*-right-invariant;*

(2) u *is* $K(\mathfrak{z}')$*-left-invariant for some finite complete subtree* $\mathfrak{z}'$*, which depends on* u*;*

(3) *for every proper complete subtree* $\mathfrak{s}\subseteq\mathfrak{z}$,

$$\int_{K(\mathfrak{s})} u(gk)\,dk = 0 \text{ for all } g\in G.$$

It is clear that $\mathcal{S}(\mathfrak{z})$ is G-left-invariant. Therefore, if $\mathcal{S}(\mathfrak{z})\neq0$ then it makes sense to consider the restriction $\lambda_{\mathfrak{z}}$ of the left regular representation to $\mathcal{S}(\mathfrak{z})$: $\lambda_{\mathfrak{z}}(h)u(g)=u(h^{-1}g)$. It is easy to see that $\rho(g)\mathcal{S}(\mathfrak{z})=\mathcal{S}(g\mathfrak{z})$ for every $g\in G$.

2. Special representations. In this section we assume again that G acts transitively on $\mathfrak{X}$ and its boundary Ω, and we classify, up to equivalence, the *special representations* of G. We shall in fact prove that there exist only two inequivalent special representations, which are the restrictions to G of the special representations of Aut($\mathfrak{X}$).

Let π be a special representation. This means according to (1.1) that the minimal tree $\mathfrak{T}$ associated to π is an edge. Let $\mathfrak{T}=e=\{a,b\}$. Define as before $K(e)=\{g\in G:\ ga=a,\ gb=b\}$, $K(\{a\})=K_a=\{g\in G:\ ga=a\}$, $K(\{b\})=K_b=\{g\in G:\ gb=b\}$,

$$P_\pi(e) = \int_{K(e)} \pi(k)\ dk \neq 0 ,$$

and

$$P_\pi(\{a\}) = \int_{K_a} \pi(k)\ dk = \int_{K_b} \pi(k)\ dk = P_\pi(\{b\}) = 0.$$

We observe now that the hypotheses on G imply that its action is *doubly transitive*. This means that we may replace the edge e with any other edge e' because, if $g\in G$ is such that $ga=a'$ and $gb=b'$, then $P_\pi(e')=P_\pi(ge)=\pi(g)P_\pi(e)\pi(g^{-1})$, and similarly

$$P_\pi(\{a'\}) = \pi(g)P_\pi(\{a\})\pi(g^{-1}), \qquad P_\pi(\{b'\}) = \pi(g)P_\pi(\{b\})\pi(g^{-1}).$$

This assertion is proved using the fact that $K(ge)=gK(e)g^{-1}$, and changing the variables in the integrals defining the projections. We now fix an edge $e_o=\{a,b\}$, and consider the space $\mathcal{S}(e_o)$.

We observe first of all that $\mathcal{S}(e_o)$ is G-left-invariant. Therefore the condition

$$\int_{K_a} u(kg)\ dk = 0$$

for elements of $\mathcal{S}(e_o)$ implies that, for every $h\in G$,

$$\int_{K_a} u(h^{-1}kg)\ dk = 0.$$

Therefore

$$\int_{K_a} u(h^{-1}khg)\ dk = 0,$$

for every h and every g. In other words, for every $h\in G$,

$$\int_{K_{ha}} u(kg)\, dk = 0.$$

Furthermore, the condition that there exists a finite complete subtree $\mathfrak{z}$ such that u is $K(\mathfrak{z})$-left-invariant may be replaced by the condition

(2) *there exists* n *such that, if* kx=x *for all vertices at distance* n *from* a, *then* u(kg)=u(g).

In other words the complete subtree $\mathfrak{z}$ may be replaced by the complete subtree $\mathfrak{X}_n=\{x: d(x,a)\le n\}$.

Since the action of G is doubly transitive on $\mathfrak{X}$, G acts *transitively* on the set of *oriented edges* [x,y] where $\{x,y\}\in\mathfrak{E}$. Since $[x,y]\ne[y,x]$ we may identify the set of oriented edges with the union $\mathfrak{E}^+\cup\mathfrak{E}^-$ of two copies of $\mathfrak{E}$. This identification may be made in such a way that whenever $[x,y]\in\mathfrak{E}^+$ (respectively $[x,y]\in\mathfrak{E}^-$) all the other q oriented edges [x,t] also belong to $\mathfrak{E}^+$ (respectively $\mathfrak{E}^-$). This is possible as shown in Fig.1.

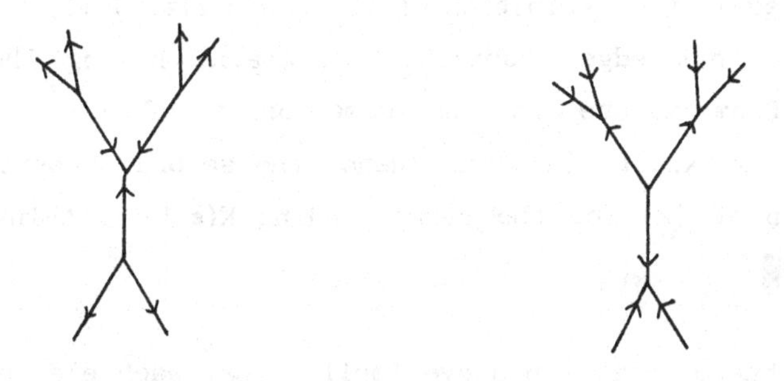

Fig.1

The stabilizer of [a,b] in $\mathfrak{E}^+\cup\mathfrak{E}^-$ is the compact group $K(e_o)$. Accordingly the map $g \longrightarrow [ga,gb]$ defines a bijection from the quotient space $G/K(e_o)$ to the set of oriented edges $\mathfrak{E}^+\cup\mathfrak{E}^-$. Therefore every $K(e_o)$-right-invariant function on G may be identified with a function on $\mathfrak{E}^+\cup\mathfrak{E}^-$. In particular $\mathcal{S}(e_o)$ may be identified with a space of functions $\tilde{\mathcal{S}}$ defined on $\mathfrak{E}^+\cup\mathfrak{E}^-$. A function u on $\mathfrak{E}^+\cup\mathfrak{E}^-$ may be identified of course with a pair of functions (u^+,u^-) defined on $\mathfrak{E}$, whose components correspond to the restrictions of u to $\mathfrak{E}^+$ and $\mathfrak{E}^-$, respectively. The space $\tilde{\mathcal{S}}$ may be characterized as follows.

(2.1) *LEMMA. Let* u *be a function on* $\mathfrak{E}^+\cup\mathfrak{E}^-$, *then* $u\in\tilde{\mathcal{S}}$ *if and only if the following conditions are satisfied.*

(1) *For every* $x\in\mathfrak{X}$ *the sum of the values of* u^+, *and separately of* u^-, *over the edges having a preassigned distance from* x, *is zero.*

(2) *There exists a positive integer* n *such that if* $\mathbf{d}(a,x)=n$ *then the restrictions of the functions* u^+ *and* u^- *to the edges contained in the cones* $\mathfrak{E}(a,x)$ *depend only on the distance of the edge from* a.

PROOF. The first property is the exact translation of property (1) for $K(e_o)$-right-invariant functions on G. For the second property recall the definition of the cones $\mathfrak{E}(a,x)$ (I,1). It is then clear that edges contained in $\mathfrak{E}(a,x)$ having the same distance from x are in the same orbit of the subgroup $K(\mathfrak{X}_n)=\{g\in G:\ gx=x,\ \text{if } \mathbf{d}(a,x)=n\}$. Hence the second property is a translation of (2) for the corresponding $K(e_o)$-right-invariant functions.∎

The characterization above implies that each element of $\tilde{\mathcal{S}}$ is uniquely determined by the values it assumes on a finite number of edges.

To be precise assume that $[a,b]\in\mathfrak{E}^+$ and suppose that $u\in\tilde{\mathcal{S}}$, and that u is $K(\mathfrak{X}_n)$-left-invariant, where $\mathfrak{X}_n=\{y:\ \mathbf{d}(y,a)\le n\}$. Let $\mathbf{d}(x,a)=n$ and let t_x be the vertex such that $\mathbf{d}(a,t_x)=n-1$ and $\mathbf{d}(t_x,x)=1$. Suppose that $u^+([t_x,x])=\alpha$. Then u^+ has the value $-\alpha/q$ on the positively oriented edges having distance n from a, and containing x; in general u^+ has the value $(-1)^{k+1}\alpha/q^{k+1}$ on the positively oriented edges contained in $\mathfrak{C}(a,x)$ and having distance n+k from a. Thus the values of u^+ outside $\mathfrak{X}_n$ are determined by the values on the positively oriented edges having distance n from a.

In particular the subspace of $\mathcal{S}(\mathfrak{e}_o)$ consisting of all $K(\mathfrak{X}_n)$-left-invariant functions is finite-dimensional. With this in mind we can prove the following theorem.

(2.2) *THEOREM. If* $u\in\tilde{\mathcal{S}}$ *then* $\sum_{\mathfrak{v}\in\mathfrak{C}}(|u^+(\mathfrak{v})|^2+|u^-(\mathfrak{v})|^2)<\infty$. *In addition, if* u *is not identically zero, then*

$$\sum_{\mathfrak{v}\in\mathfrak{C}}(|u^+(\mathfrak{v})|+|u^-(\mathfrak{v})|)=\infty.$$

Therefore the space $\mathcal{S}(\mathfrak{e}_o)$ *is contained in* $L^2(G)$ *and* $\mathcal{S}(\mathfrak{e}_o)\cap L^1(G)=\{0\}$.

PROOF. Assume that u is $K(\mathfrak{X}_n)$-left-invariant. Then

$$\sum_{\mathfrak{v}\subseteq\mathfrak{C}(a,x)}|u^+(\mathfrak{v})|^2=\sum_{n=0}^{\infty}\sum_{\substack{\mathfrak{v}\subseteq\mathfrak{C}(a,x)\\ d(\mathfrak{v},a)=n}}|u^+(\mathfrak{v})|^2$$

$$=\sum_{n=0}^{\infty}\frac{|u([t_x,x])|^2\ q^{k+1}}{q^{2(k+1)}}=|u^+([t_x,x])|^2\ \frac{q}{q-1}\ .$$

Therefore,

$$\sum|u^+(\mathfrak{v})|^2=\sum_{\mathfrak{v}\subseteq\mathfrak{X}n}|u^+(\mathfrak{v})|^2+\frac{q}{q-1}\sum_{d(\mathfrak{v},x)=n}|u^+([t_x,x])|^2<\infty.$$

The same calculation is true for u^-. We have thus proved the first part of the theorem. Let now $u^+\neq0$, and suppose that u^+ is $K(\mathfrak{X}_n)$-left-invariant. It is easy to see (using the same notation as in the first part of the proof) that if $u^+([t_x,x])=0$, for all vertices x such that $d(x,a)=n$, then $u^+(y)=0$, for all $y\in\mathfrak{X}_n$. We may assume therefore that for some vertex x, with $d(x,a)=n$, $u^+([t_x,x])=\alpha\neq0$. Then

$$\sum_{\mathfrak{v}\subseteq\mathfrak{C}(a,x)}|u^+(\mathfrak{v})|=\sum_{k=0}^{\infty}(|\alpha|/q^{k+1})\ q^{k+1}=+\infty\ .$$

The L^2-norm of an element of $\mathcal{S}(e_o)$ is a multiple of the sum

$$\left[\sum_{\mathfrak{v}\subseteq\mathfrak{E}} |u^+(\mathfrak{v})|^2 + \sum_{\mathfrak{v}\subseteq\mathfrak{E}} |u^-(\mathfrak{v})|^2\right]^{1/2},$$

where u^+ and u^- are the restrictions to $\mathfrak{E}^+$ and $\mathfrak{E}^-$, respectively, of the function of $\tilde{\mathcal{S}}$ corresponding to that element. A similar statement is true for the L^1-norm and the theorem follows.∎

(2.3) *PROPOSITION* *The subspace of* $\mathcal{S}(e_o)$ *consisting of all* $K(e_o)$*-left-invariant functions is a two-dimensional subspace. For every* $K(e_o)$*-left-invariant* $f\in\mathcal{S}(e_o)$ *the following equality holds:*

$$\|f\|_2^2 = (m(K_a)/(q-1))\ (|f^+(e_o)|^2 + |f^-(e_o)|^2).$$

PROOF. Let M be the subspace of $\mathcal{S}(e_o)$ consisting of $K(e_o)$-left-invariant functions. If $f\in M$ then f is constant on the edges at distance n from e_o. We shall presently show that there exist exactly two linearly independent elements of $\mathcal{S}(e_o)$ which are $K(e_o)$-invariant. Indeed if $u\in\mathcal{S}(e_o)$ is $K(e_o)$-invariant, with $e_o=\{a,b\}$ and $[a,b]\in\mathfrak{E}^+$, and if $u^+(e_o)=u([a,b])=\alpha$, $u^-(e_o)=u([b,a])=\beta$, then u^+ takes the value $-\alpha/q$ on the q positively oriented edges starting with b, and on the q positively oriented edges ending with a. The other values of u^+ may be inductively determined. Similarly the values of u^- may be determined from the value of u^- on e_o. Let v_1 be the function which is zero on $\mathfrak{E}^-$ and determined as above on $\mathfrak{E}^+$, starting with the value $v_1([a,b])=1$, and let v_2 be zero on $\mathfrak{E}^+$ and similarly determined on $\mathfrak{E}^-$, starting with the value $v_2([a,b])=1$. Then $u=\alpha v_1+\beta v_2$. This shows that v_1 and v_2 form a basis for the space of $K(e_o)$-invariant functions of $\mathcal{S}(e_o)$. Therefore the map $f\longrightarrow(f^+(e_o),f^-(e_o))$ is a linear isomorphism of M onto $\mathbb{C}^2$. We have that

$$\|f\|_2^2 = m(K(e_o))\Big(\sum_{\mathfrak{v}\subseteq\mathfrak{E}} (|f^+(\mathfrak{v})|^2+|f^-(\mathfrak{v})|^2)\Big);$$

as in the proof of (2.2), we can prove, easily, that

$$\sum_{\mathfrak{v}\subseteq\mathfrak{E}} |f^{+}(\mathfrak{v})|^2 = |f^{+}(e_o)|^2((q+1)/(q-1))$$

and

$$\sum_{\mathfrak{v}\subseteq\mathfrak{E}} |f^{-}(\mathfrak{v})|^2 = |f^{-}(e_o)|^2((q+1)/(q-1)).$$

G acts transitively on Ω and $K(e_o)$ has index q+1 in K_a; therefore $m(K_a)=(q+1)m(K(e_o))$ and the proposition follows.∎

Theorem (2.2) and Proposition (2.3) imply that $\mathcal{G}(e_o)$ is a nontrivial left-invariant subspace of $L^2(G)$. Let $\mathcal{M}(e_o)$ be the closure of $\mathcal{G}(e_o)$ in $L^2(G)$. Let λ_{e_o} be the subrepresentation of the left regular representation of G on $L^2(G)$ which corresponds to the invariant subspace $\mathcal{M}(e_o)$. Since $\rho(g)\mathcal{G}(e_o)=\mathcal{G}(ge_o)$, we have that $\rho(g)\mathcal{M}(e_o)=\mathcal{M}(ge_o)$; therefore λ_{e_o} and λ_{ge_o} are unitarily equivalent. The space $\mathcal{M}(e_o)$ can be regarded as the subspace of $\ell^2(\mathfrak{E}^+\cup\mathfrak{E}^-)$ consisting of functions with the property (1) of Lemma (2.1). To prove this, it is enough to see that every function $f\in\ell^2(\mathfrak{E}^+\cup\mathfrak{E}^-)$ with the property (1) of Lemma (2.1) is a limit in ℓ^2 of functions of $\mathcal{G}(e_o)$. The proof is straightforward: if u_n is the function in $\mathcal{G}(e_o)$ such that u_n is constant on the orbit of $K(\mathfrak{X}_n)=\{h\in G : hx=x$ for every x such that $d(a,x)=n\}$ and $u_n=f$ on $\mathfrak{X}_n$, then it is easy to see, as in the proof of Theorem (2.3), that

$$\|f-u_n\|_2^2 \le m(K(e_o))(1+(q-1)^{-1/2})^2(\sum_{\mathfrak{v}\not\subseteq\mathfrak{X}_{n-1}} |f^{+}(\mathfrak{v})|^2+|f^{-}(\mathfrak{v})|^2)$$

because

$$\sum_{\mathfrak{v}\not\subseteq\mathfrak{X}_n} (|u_n^{+}(\mathfrak{v})|^2+|u_n^{-}(\mathfrak{v})|^2)$$

$$= (q-1)^{-1} \sum_{\mathfrak{v}\subseteq\mathfrak{X}_n/\mathfrak{X}_{n-1}} (|f^{+}(\mathfrak{v})|^2+|f^{-}(\mathfrak{v})|^2).$$

Since $f\in\ell^2(\mathfrak{E}^+\cup\mathfrak{E}^-)$, for every $\varepsilon>0$ there exists n such that $\|f-u_n\|_2<\varepsilon$. As in Theorem (2.2), we have that $\mathcal{M}(e_o)\cap L^1(G)=\{0\}$. It is also clear that every $K(e_o)$-left-invariant function of $\mathcal{M}(e_o)$ is in $\mathcal{S}(e_o)$; therefore Proposition (2.3) is true for $\mathcal{M}(e_o)$. This implies that the range of the projection $P_\lambda(e_o)$ on $\mathcal{M}(e_o)$, that is the subspace $P_\lambda(e_o)(\mathcal{M}(e_o))$, is two-dimensional. Since $K(e_o)$ has index 2 in the stabilizer $\tilde{K}(e_o)=\{g\in G:\ ge_o=e_o\}$, $g_oK(e_o)=K(e_o)g_o$ for every inversion g_o on e_o. In particular, for every $f\in\mathcal{M}(e_o)$, $P_\lambda(e_o)f(1_G)=f(1_G)$ (1_G is the identity of G) and $P_\lambda(e_o)f(g_o)=f(g_o)$. This means that the function $P_\lambda(e_o)f$ is the $K(e_o)$-left-invariant function of $\mathcal{S}(e_o)$ which corresponds to the pair $(f^+(e_o),f^-(e_o))$ for every $f\in\mathcal{M}(e_o)$, and so $f\in\mathcal{M}(e_o)$ is orthogonal in $L^2(G)$ to $P_\lambda(e_o)(\mathcal{M}(e_o))$ if and only if $f^+(e_o)=f^-(e_o)=0$, that is f is identically zero on $\tilde{K}(e_o)$.

(2.4) *LEMMA. Every nontrivial closed left-invariant subspace of $\mathcal{M}(e_o)$ contains a nontrivial $K(e_o)$-left-invariant function of $\mathcal{M}(e_o)$.*

PROOF. Let H be a nontrivial closed subspace of $\mathcal{M}(e_o)$, invariant under left translation. Let $u\in H$, $u\neq0$; we may also suppose that $u(1_G)\neq0$, replacing u by one of its left translates. Since H is left-invariant, $P_\lambda(e_o)H\subset H$ and $P_\lambda(e_o)u(1_G)=u(1_G)\neq0$ because u is $K(e_o)$-right-invariant, and so u is constant on $K(e_o)$. This proves the lemma.∎

Lemma (2.4) implies that e_o is a minimal tree for λ_{e_o} and for every subrepresentation of λ_{e_o}. In particular every irreducible subrepresentation of λ_{e_o} is special. Since the space of $K(e_o)$-left-invariant functions of $\mathcal{M}(e_o)$ is two-dimensional, Lemma (2.4) and Proposition (2.3) imply that

λ_{e_o} either is irreducible or is the sum of two irreducible subrepresentations.

(2.5) *LEMMA. Every unitary special irreducible representation of G is unitarily equivalent to a subrepresentation of* λ_{e_o}.

PROOF. Let π be a unitary special irreducible representation of G and $\xi\in\mathcal{H}_\pi$ a nontrivial $K(e_o)$-invariant vector. $(\pi(g)\xi,\xi)$ is a nontrivial $K(e_o)$-bi-invariant function of $\mathcal{S}(e_o)$; in particular $(\pi(g)\xi,\xi)\in L^2(G)$. This implies that π is a square-integrable representation [D1, 14.1.3]. Since G is unimodular, every coefficient of π is in $L^2(G)$; therefore $(\pi(g)\xi,\eta)\in\mathcal{M}(e_o)$ for every $\eta\in\mathcal{H}_\pi$. By [D1, 14.3.3] $\|(\pi(.)\xi,\eta)\|_2=\|\xi\|\,\|\eta\|/(\sqrt{d_\pi})$, where d_π is the formal dimension of π. Hence the operator $U\colon \mathcal{H}_\pi\longrightarrow\mathcal{M}(e_o)$, defined by $U(\eta)(.)=(\sqrt{d_\pi})\|\xi\|^{-1}(\eta,\pi(.)\xi)$, is a unitary operator intertwining the representations π and λ. The lemma follows.∎

(2.6) *THEOREM. The representation* λ_{e_o} *is the orthogonal sum of two inequivalent irreducible subrepresentations* σ_1 *and* σ_2. *The representations* σ_1 *and* σ_2 *are* L^2 *but not* L^1*-representations, and their formal dimensions are* $d_{\sigma_1}=d_{\sigma_2}=(q-1)/(2m(K_a))$.

PROOF. Let f_o be the unique $K(e_o)$-left-invariant function of $\mathcal{S}(e_o)$ such that $f_o^+(e_o)=f_o^-(e_o)=1$ and let h_o be the $K(e_o)$-left-invariant function of $\mathcal{S}(e_o)$ such that $h_o^+(e_o)=1$ and $h_o^-(e_o)=-1$. The vectors (1,1) and (1,-1) are orthogonal in $\mathbb{C}^2$; therefore, by Proposition (2.3), $(f_o,h_o)=0$ in $L^2(G)$ and $\{f_o,h_o\}$ is a basis of the space of $K(e_o)$-left-invariant functions of $\mathcal{M}(e_o)$. Let g_o be an inversion on the edge e_o; since $f_o^+=f_o^-$, $\lambda(g_o)f_o=f_o$ and $\lambda(g_o)h_o=-h_o$ because $h_o^+=-h_o^-$ and an inversion changes the orientation of every edge. Let M_1 be the closed subspace of $\mathcal{M}(e_o)$ generated by the set $\{\lambda(s)f_o\colon s\in G\}$ and let M_2 be the closed subspace of $\mathcal{M}(e_o)$ generated by the set $\{\lambda(s)h_o: s\in G\}$. The subspaces M_1 and M_2 are closed and nontrivial. Let σ_1 and σ_2 be the subrepresentations of λ_{e_o} corresponding to M_1 and M_2, respectively. Let now $F(t)=(\lambda(t)f_o,h_o)$; F is a $K(e_o)$-left-invariant function of $\mathcal{S}(e_o)$ and $F^+(e_o)=F^-(e_o)=(f_o,h_o)=0$; therefore, by Proposition (2.3), F is identically zero. This implies that $(v,w)=0$ for every $v\in M_1$ and for every $w\in M_2$. This means that the closed nontrivial invariant subspace $M_1\oplus M_2$ is the orthogonal sum of M_1 and M_2; every $K(e_o)$-left-invariant function of $\mathcal{M}(e_o)$ is contained in $M_1\oplus M_2$ because this space contains the basis $\{f_o,h_o\}$; therefore Proposition (2.3) and Lemma (2.4) imply that the orthogonal complement of $M_1\oplus M_2$ must be zero. This means that $M_1\oplus M_2=\mathcal{M}(e_o)$ and $\lambda_{e_o}=\sigma_1\oplus\sigma_2$ is the orthogonal sum of σ_1 and σ_2. Moreover, by Proposition (2.3) and Lemma (2.4), as observed, there are at most two irreducible subrepresentations of λ_{e_o}, and so σ_1 and σ_2 are irreducible special representations. We show now that σ_1 and σ_2 are inequivalent. Suppose that, on the contrary, there exists a unitary operator U mapping M_1 onto M_2 and such that $\lambda_{e_o}(g)U=U\lambda_{e_o}(g)$ for every $g\in G$. This implies that $P_\lambda(e_o)U=UP_\lambda(e_o)$ and so U takes $K(e_o)$-left-invariant functions of M_1 into $K(e_o)$-left-invariant functions of M_2. But $K(e_o)$-left-invariant functions form one-dimensional subspaces of M_1 and M_2, respectively. Therefore $Uf_o=ch_o$ for some constant c. In particular $\lambda(g_o)Uf_o=-Uf_o$ because $\lambda(g_o)h_o=-h_o$ (recall that g_o is an inversion on the edge e_o). Finally, $Uf_o=U\lambda(g_o)f_o=\lambda(g_o)Uf_o=-Uf_o$, that is $Uf_o=0$. This contradicts the injectivity of U and we conclude that the restrictions of λ_{e_o} to M_1 and M_2 are inequivalent. We recall that d_π the formal dimension of an L^2-representation is equal to $\|(\pi(.)\xi,\xi)\|_2^{-2}$ where $\|\xi\|=1$ [D1, 14.3.3]. In particular $d_{\sigma_1}=\|F\|_2^{-2}$ where $F(t)=$

$(\lambda(t)f_o,f_o)/\|f_o\|_2^2$. The function F is $K(e_o)$-left-invariant and $F\in\mathcal{M}(e_o)$ moreover $F^+(e_o)=F^-(e_o)=1$ this means that $F(t)=f_o(t)$ for every $t\in G$. Hence $d_{\sigma_1}=\|f_o\|_2^{-2}$. The fact that $d_{\sigma_2}=\|h_o\|_2^{-2}$ is similarly proved. Proposition (2.3) implies that $d_{\sigma_1}=d_{\sigma_2}=(q-1)/2m(K_a)$. The fact that σ_1 (the proof for σ_2 is similar) is not an L^1-representation is a consequence of the fact that $\mathcal{M}(e_o)\cap L^1(G)=0$. Indeed, let $(\lambda(.)\xi,\xi)\in L^1(G)$ for $\xi\in M_1$ and $\xi\neq0$; as we may replace ξ by one of its left translates we may assume that $\xi(1_G)\neq0$ (G is unimodular and hence the function $(\lambda(.)\lambda(t)\xi,\lambda(t)\xi)\in L^1(G)$ for every $t\in G$). We have that $P_\lambda(e_o)\xi(1_G)=\xi(1_G)\neq0$ and $P_\lambda(e_o)\xi\in M_1$; since $P_\lambda(e_o)\xi$ is $K(e_o)$-left-invariant, $P_\lambda(e_o)\xi=\xi(1_G)f_o$. An easy computation shows that

$$\int_G(\lambda(g)P_\lambda(e)\xi,P_\lambda(e)\xi)dg = \int_G (\lambda(g)\xi,\xi)\, dg.$$

Since $f_o\geq0$, we have that $(\lambda(.)f_o,f_o)\in L^1(G)\cap\mathcal{M}(e_o)$. This is a contradiction.∎

Theorem (2.6) and Lemma (2.5) imply that, for any closed subgroup of $\mathrm{Aut}(\mathfrak{X})$ acting transitively on $\mathfrak{X}$ and Ω, there exist only two inequivalent special irreducible unitary representations.

A similar statement is true for a closed noncompact subgroup G of $\mathrm{Aut}(\mathfrak{X})$ acting transitively on Ω but not on $\mathfrak{X}$, that is a subgroup G which has exactly the two orbits $\mathfrak{X}^+$ and $\mathfrak{X}^-$ (see Proposition (10.2) of Chapter **I**); for instance $\mathrm{Aut}(\mathfrak{X})^+$ or $\mathbf{PSL}(2,\mathfrak{F})$ where $\mathfrak{F}$ is a local field. In this case $G/K(e_o)$ can be regarded as the set of nonoriented edges. This implies that the subspace of $\mathcal{M}(e_o)$ consisting of all $K(e_o)$-left-invariant functions is one-dimensional and λ_{e_o} is irreducible. Proposition (2.3) and Lemma (2.4) are true with the same proof.

Hence, for such a G, there exists only one unitary special irreducible representation. This is an L^2- but not L^1-representation and its formal dimension is equal to $(q-1)/m(K_a)$.

3. Cuspidal representations and the Plancherel formula of Aut($\mathfrak{X}$). In this section we suppose that $G=\mathrm{Aut}(\mathfrak{X})$. The results we present here are not always true for groups acting transitively on $\mathfrak{X}$ and on Ω. In particular (3.2) fails for **PGL**$(2,\mathfrak{F})$ thought of as a group of automorphisms of its canonical tree. It will be clear, however, from the proof of (3.1), that all the results of this section remain true for a closed subgroup of $\mathrm{Aut}(\mathfrak{X})$ containing sufficiently many rotations; in particular they remain true for $\mathrm{Aut}(\mathfrak{X})^+$, the group generated by *all* rotations.

(3.1) *LEMMA. Let* $\mathfrak{t}$ *be a complete subtree of* $\mathfrak{X}$ *with* $\mathrm{diam}(\mathfrak{t})\geq 2$. *Let* $\mathfrak{y}$ *be a complete subtree not containing* $\mathfrak{t}$. *Then there exists a proper complete subtree* $\mathfrak{z}\subset\mathfrak{t}$ *such that* $K(\mathfrak{z})\subseteq K(\mathfrak{y})K(\mathfrak{t})$.

PROOF. We first show that we may assume that $\mathfrak{t}\cap\mathfrak{y}$ contains an edge. Indeed, if $\mathfrak{t}\cap\mathfrak{y}$ is empty or consists of only one vertex, then there exists a unique vertex $x\in\mathfrak{t}$ of minimal distance from $\mathfrak{y}$. For if m is the minimal distance of $\mathfrak{t}$ and $\mathfrak{y}$ and $m=\mathbf{d}(x,y)$ $=\mathbf{d}(x',y')$ with $x,x'\in\mathfrak{t}$, $y,y'\in\mathfrak{y}$ and $x\neq x'$, then $[x,y]$ and $[x',y']$ intersect $\mathfrak{t}$ or $\mathfrak{y}$ only at the end points. But $[x,x']\subseteq\mathfrak{t}$ and $[y,y']\subseteq\mathfrak{y}$. This means that $[x,x']\cup[x',y']\cup[y',y]\cup[y,x]$ is a circuit: a contradiction. Let $\mathfrak{y}'=\{x\in\mathfrak{X}:\ \mathbf{d}(x,\mathfrak{y})\leq m+1\}$. Then $\mathfrak{y}'$ is a complete subtree and $\mathfrak{y}'\cap\mathfrak{t}$ contains an edge. Observe that $\mathfrak{y}'$ does not contain $\mathfrak{t}$. Indeed, if all the vertices of $\mathfrak{t}$ had distance at most m+1 from $\mathfrak{y}$, then all vertices of $\mathfrak{t}$ would have distance at most 1 from the vertex closest to $\mathfrak{y}$, which would imply that $\mathfrak{t}$ has diameter less than 2 or that $\mathfrak{t}$ is not complete. Clearly $K(\mathfrak{y}')\subseteq K(\mathfrak{y})$. Therefore substituting, if needed, $\mathfrak{y}'$ for $\mathfrak{y}$ we may suppose that $\mathfrak{y}\cap\mathfrak{x}$ contains an edge. Let $\mathfrak{z}=\mathfrak{y}\cap\mathfrak{x}$ then $\mathfrak{z}$ is a complete subtree of $\mathfrak{x}$ different from $\mathfrak{x}$ itself. We prove that $K(\mathfrak{z})\subseteq K(\mathfrak{y})K(\mathfrak{x})$. Let $x_1,\dots,x_s$ be the vertices of $\mathfrak{z}$ which have degree 1. Let $\mathfrak{C}_i=\mathfrak{C}(x,x_i)$, with $x\in\mathfrak{z}$, be the cones of vertex x_i which intersect $\mathfrak{z}$ only in $\{x_i\}$. Let K_i be the group of rotations which fix every element outside $\mathfrak{C}_i$. Clearly $K_i\subseteq K(\mathfrak{z})$, and in particular every element of K_i stabilizes x_i. In addition every element $k\in K(\mathfrak{z})$ may be written as a product $k=k_1\dots k_s$, with $k_i\in K_i$ and $k_ik_j=k_jk_i$ for $i\neq j$. Therefore $K(\mathfrak{z})$ is the direct product of the groups $K_1,\dots,K_s$. We prove now that, for every $i=1,..,s$, either $K_i\subseteq K(\mathfrak{x})$ or $K_i\subseteq K(\mathfrak{y})$. Indeed $K_i\subseteq K(\mathfrak{x})$ if and only if $\mathfrak{C}_i\setminus\{x_i\}$ does not intersect $\mathfrak{x}$. Similarly $K_i\subseteq K(\mathfrak{y})$ if and only if $\mathfrak{C}_i\setminus\{x_i\}$ does not intersect $\mathfrak{y}$. Therefore, if K_i is contained neither in $K(\mathfrak{x})$ nor in $K(\mathfrak{y})$, then $\mathfrak{C}_i\setminus\{x_i\}$ must contain vertices of both $\mathfrak{x}$ and $\mathfrak{y}$. Let $x\in\mathfrak{x}\cap(\mathfrak{C}_i\setminus\{x_i\})$ and $y\in\mathfrak{y}\cap(\mathfrak{C}_i\setminus\{x_i\})$. Then $[x,x_i]\subseteq\mathfrak{x}\cap\mathfrak{C}_i$ and $[y,x_i]\subseteq\mathfrak{y}\cap\mathfrak{C}_i$. Therefore there exist vertices $x'\in\mathfrak{x}\cap\mathfrak{C}_i$ and $y'\in\mathfrak{y}\cap\mathfrak{C}_i$, such that $\mathbf{d}(x',x_i)=\mathbf{d}(y',x_i)=1$. Since $\mathfrak{x}$, $\mathfrak{y}$ and $\mathfrak{x}\cap\mathfrak{y}$ are complete subtrees, this implies that $x',y'\in\mathfrak{x}\cap\mathfrak{y}$, a contradiction since we assumed that the degree of x_i is 1. We have thus proved that, for each i, $K_i\subseteq K(\mathfrak{x})$ or $K_i\subseteq K(\mathfrak{y})$. If $k\in K(\mathfrak{z})$ then $k=k_1\dots k_s$ with $k_i\in K_i$. Since the k_i commute, and each belongs to either $K(\mathfrak{x})$ or $K(\mathfrak{y})$, we may write $k=k'k''$, with $k'\in K(\mathfrak{x})$, and $k''\in K(\mathfrak{y})$. In other words $K(\mathfrak{z})\subseteq K(\mathfrak{x})K(\mathfrak{y})$. ∎

(3.2) *PROPOSITION. Let* $\mathfrak{x}$ *be a finite complete subtree of* $\mathfrak{X}$ *and suppose that the maximum distance of two vertices of* $\mathfrak{x}$ *is at least 2, in symbols* $\mathrm{diam}(\mathfrak{x})\geq 2$. *Let* $\mathfrak{y}\subseteq\mathfrak{X}$ *be a complete subtree. If* u *is a* $K(\mathfrak{y})$*-left-invariant element of* $\mathcal{S}(\mathfrak{x})$, *then*

$$\mathrm{supp}(u)\subseteq\{g\in\mathrm{Aut}(\mathfrak{X}):\ g\mathfrak{x}\subseteq\mathfrak{y}\}.$$

PROOF. We must prove that, if $u\in\mathcal{S}(\mathfrak{x})$ and u is $K(\mathfrak{y})$-invariant, then $u(g)=0$, for all automorphisms g such that $g\mathfrak{x}\not\subseteq\mathfrak{y}$. According to (3.1) this is true for $g=e$. Indeed if $\mathfrak{y}$ does not contain $\mathfrak{x}$

there exists a complete subtree $\mathfrak{z}$ properly contained in $\mathfrak{T}$ such that $K(\mathfrak{z})\subseteq K(\mathfrak{T})K(\mathfrak{y})$. Since u is $K(\mathfrak{T})$-right-invariant and $K(\mathfrak{y})$-left-invariant u must be constant on $K(\mathfrak{z})$. But, since $\mathfrak{z}$ is properly contained in $\mathfrak{T}$, the average of u on $K(\mathfrak{z})$ must be zero, which implies that u is zero on $K(\mathfrak{z})$. In particular $u(e)=0$. Let now $g\in\mathrm{Aut}(\mathfrak{X})$ be such that $\mathfrak{y}$ does not contain $g\mathfrak{T}$; then $\rho(g)u\in\mathscr{S}(g\mathfrak{T})$ and $\rho(g)u$ is $K(\mathfrak{y})$-left-invariant. We also have that $g\mathfrak{T}$ is a complete subtree of diameter at least 2. This means that $\rho(g)u(e)=u(g)=0$. ∎

(3.3) *COROLLARY. If $\mathfrak{T}$ is a finite complete subtreee with diameter at least 2, then every element of $\mathscr{S}(\mathfrak{T})$ has compact support. In particular every cuspidal irreducible representation of* $\mathrm{Aut}(\mathfrak{X})$ *is an L^1-representation.*

PROOF. If $u\in\mathscr{S}(\mathfrak{T})$ then there exists $\mathfrak{y}$ such that u is $K(\mathfrak{y})$-left-invariant. This means by (3.2) that the support of u is contained in the compact set $\{g\in\mathrm{Aut}(\mathfrak{X}):\ g\mathfrak{T}\subseteq\mathfrak{y}\}$. If π is cuspidal and ξ is a nontrivial $K(\mathfrak{T})$-invariant vector, where $\mathfrak{T}$ is a minimal tree for π, then $(\pi(.)\xi,\xi)$ is in $\mathscr{S}(\mathfrak{T})$ and π is an L^1-representation. ∎

(3.4) *COROLLARY. Let π be a cuspidal irreducible representation of* $\mathrm{Aut}(\mathfrak{X})$ *with minimal tree $\mathfrak{T}$. Let $[\mathfrak{T}]=\{g\mathfrak{T}:$ $g\in\mathrm{Aut}(\mathfrak{X})\}$; then $\mathfrak{T}'$ is a minimal tree of π iff $\mathfrak{T}'\in[\mathfrak{T}]$.*

PROOF. As observed for special representations, if $\mathfrak{T}$ is a minimal tree of π then also $g\mathfrak{T}$ is a minimal tree for every $g\in\mathrm{Aut}(\mathfrak{X})$ because $\mathrm{card}(g\mathfrak{T})=\mathrm{card}(\mathfrak{T})$ and $\pi(g)P_\pi(\mathfrak{T})\pi(g)^{-1}=P_\pi(g\mathfrak{T})$. Conversely if $\mathfrak{T}'$ is a minimal tree of π then $\mathrm{card}(\mathfrak{T})=\mathrm{card}(\mathfrak{T}')=\ell_\pi$; let ξ be a nontrivial $K(\mathfrak{T})$-invariant vector and η be a nontrivial $K(\mathfrak{T}')$-invariant vector. Then $(\pi(.)\xi,\eta)$ is a $K(\mathfrak{T}')$-left-invariant nontrivial function of $\mathscr{S}(\mathfrak{T})$. Proposition (3.2) implies that the support of $(\pi(.)\xi,\eta)$ is contained in the set $\{g\in\mathrm{Aut}(\mathfrak{X}):\ g\mathfrak{T}\subset\mathfrak{T}'\}$ which is nonempty because $(\pi(.)\xi,\eta)$ is nontrivial. Therefore there exists $g\in\mathrm{Aut}(\mathfrak{X})$ such that $g\mathfrak{T}\subset\mathfrak{T}'$;

this means that $g\mathfrak{Z}=\mathfrak{Z}'$ because card($g\mathfrak{Z}$)=card($\mathfrak{Z}$)=card($\mathfrak{Z}'$). ∎

Assume now, as before, that $\mathfrak{Z}$ is a finite complete subtree with diam($\mathfrak{Z}$)≥2. Let $\tilde{K}(\mathfrak{Z})=\{g\in Aut(\mathfrak{Z}):g\mathfrak{Z}=\mathfrak{Z}\}$. Then $K(\mathfrak{Z})$ is a normal subgroup of $\tilde{K}(\mathfrak{Z})$ and the finite group $\tilde{K}(\mathfrak{Z})/K(\mathfrak{Z})$ is isomorphic to the group of automorphisms of the finite tree $\mathfrak{Z}$. It is interesting to observe that every function of $\mathcal{S}(\mathfrak{Z})$ which is $K(\mathfrak{Z})$-left-invariant is supported on $\tilde{K}(\mathfrak{Z})$ and may be identified with a function on $\tilde{K}(\mathfrak{Z})/K(\mathfrak{Z})$. Similarly an element of $\mathcal{S}(\mathfrak{Z})$ which is $K(\mathfrak{y})$-left-invariant may be identified with a function on the cosets $gK(\mathfrak{Z})$ which is zero on the cosets which do not intersect the compact set $\{g:g\mathfrak{Z}\subseteq\mathfrak{y}\}$. Since only finitely many cosets intersect a compact set we conclude that the space of $K(\mathfrak{y})$-left-invariant elements of $\mathcal{S}(\mathfrak{Z})$ is finite-dimensional.

Before going on with the study of cuspidal representations we want to show that for every complete finite tree $\mathfrak{Z}$ there exists at least one cuspidal representation having $\mathfrak{Z}$ as its minimal tree. In other words we shall prove that $\mathcal{S}(\mathfrak{Z})\neq\{0\}$. This will be accomplished in several steps. We first need a few observations.

Let $\mathfrak{Z}'$ be a complete proper maximal subtree of $\mathfrak{Z}$. It follows that $\mathfrak{Z}\setminus\mathfrak{Z}'\subset\partial\mathfrak{Z}$ and card($\mathfrak{Z}\setminus\mathfrak{Z}'$)=q. If diam($\mathfrak{Z}$)=2, that is $\mathfrak{Z}=V_1(\{x_o\})$ for some $x_o\in\mathfrak{X}$, then the maximal proper complete subtrees of $\mathfrak{Z}$ are the q+1 edges of $\mathfrak{Z}$. If diam($\mathfrak{Z}$)>2 then the maximal proper complete subtrees of $\mathfrak{Z}$ correspond, bijectively, to the vertices of $\partial\mathfrak{Z}^o$; in fact if $v\in\partial\mathfrak{Z}^o$ then $\mathfrak{Z}\setminus\{a_1,a_2,\dots,a_q\}$ is a complete maximal subtree, where $a_1,a_2,\dots,a_q$ are the q vertices of $\partial\mathfrak{Z}$ at distance 1 from v. If $\mathfrak{Z}'$ is maximal proper and $g\in\tilde{K}(\mathfrak{Z})$ then also $g\mathfrak{Z}'$ is a maximal proper complete subtree, in fact $g\mathfrak{Z}^o=\mathfrak{Z}^o$. This proves that $K(\mathfrak{Z}^o)$ is normal in $\tilde{K}(\mathfrak{Z})$ and that the inner automorphisms of $\tilde{K}(\mathfrak{Z})$ permute the groups $K(\mathfrak{Z}_1),K(\mathfrak{Z}_2),\dots,K(\mathfrak{Z}_j)$ where $\mathfrak{Z}_1,\mathfrak{Z}_2,\dots,\mathfrak{Z}_j$ are the complete maximal proper subtrees of $\mathfrak{Z}$. Also it is clear that

$K(\mathfrak{J}) \subset K(\mathfrak{J}_i) \subset \tilde{K}(\mathfrak{J})$ for every $i=1,2,\ldots,j$.

(3.5) *DEFINITION. A unitary representation of* $\tilde{K}(\mathfrak{J})$ *is called standard if it has no nonzero* $K(\mathfrak{J}_i)$*-invariant vectors for every* $i=1,2,\ldots,j$. *Let* $(\tilde{K}(\mathfrak{J}))^{\wedge}_{o}$ *be the set of all unitary irreducible standard representations of* $\tilde{K}(\mathfrak{J})$ *which are trivial on* $K(\mathfrak{J})$.

The finite group $\tilde{K}(\mathfrak{J})/K(\mathfrak{J})$ is isomorphic to $Aut(\mathfrak{J})$, the group of all isometries of the finite tree $\mathfrak{J}$, and $K(\mathfrak{J}_i)/K(\mathfrak{J})$ is isomorphic to the group $\{h \in Aut(\mathfrak{J}) : h(x)=x \text{ for every } x \in \mathfrak{J}_i\}$. Therefore $(\tilde{K}(\mathfrak{J}))^{\wedge}_{o}$ can be regarded as the finite set of all irreducible standard unitary representations of the finite group $Aut(\mathfrak{J})$. We prove now that $(\tilde{K}(\mathfrak{J}))^{\wedge}_{o} \neq \emptyset$ for every complete finite subtree $\mathfrak{J}$ of $\mathfrak{X}$ with $diam(\mathfrak{J}) \geq 2$.

(3.6) *LEMMA. Let* G *be a finite group and* $H_1, H_2, \ldots, H_j$ *be* j *subgroups of* G. *Then there exists a (unitary) irreducible representation* π *of* G, *such that, for every* $i=1,2,\ldots,j$, π *has no nontrivial* H_i*-invariant vectors if and only if there exists a function* f *on* G *not identically zero such that* $\sum_{h \in H_i} f(ght)=0$ *for every* $g,t \in G$ *and* $i=1,2,\ldots,j$.

PROOF. Let M be the space of functions f on G such that $\sum_{h \in H_i} f(ght)=0$ for every $g,t \in G$ and $i=1,2,\ldots,j$. M is a bi-invariant finite-dimensional space of functions on G. Hence M is a finite sum of irreducible invariant subspaces. It is clear that if f is a H_i-left-invariant function of M then $f \equiv 0$. This proves that if $M \neq 0$ then there exist irreducible unitary representations of G without nonzero H_i-invariant vectors for every $i=1,2,\ldots,j$. Conversely, if π is such a representation, then π is equivalent to a subrepresentation of λ_G; let N be the left-invariant space of functions on G which corresponds to π.

$P_\lambda(H_i)= |H_i|^{-1}\sum_{h\in H_i} \lambda(h)$ is the projection on the space consisting of H_i-left-invariant function. Since π has no nonzero H_i-invariant vectors it follows that $P_\lambda(H_i)f\equiv 0$ for every $f\in N$. Since N is left-invariant, $P_\lambda(H_i)\lambda(g)f\equiv 0$ for every $g\in G$, $f\in N$ and $i=1,2,\ldots j$. Thus $N\subset M$ and so $M\neq 0$. This proves the lemma. ∎

(3.7) *LEMMA. Let* G *be a finite group and* $H=H_1\times H_2\times\ldots\times H_j\leq G$ *the direct product of* j *nontrivial subgroups of* G. *If the inner automorphisms of* G *permute the subgroups* $H_1,H_2,\ldots,H_j$ *then there exists an irreducible (unitary) representation* π *of* G *such that, for every* $i=1,2,\ldots,j$, π *has no nonzero* H_i*-invariant vectors.*

PROOF. First, we prove the lemma in the special case H=G; by Lemma (3.6) it is enough to prove that there exists a function $f\neq 0$ on $H=H_1\times H_2\times\ldots\times H_j$ such that $\sum_{h\in H_i} f(ht)=0$ for every $t\in G$ and $i=1,2,\ldots,j$ because, for every i H_i is a normal subgroup of H and so $gH_it=gtH_i$. Let $E_i=\{x_i,y_i\}$ where $x_i,y_i\in H_i$ and $x_i\neq y_i$ for every i. Let $E=E_1\times E_2\times\ldots\times E_j$ and for $v\in E$ let N(v) be the number of x_i occurring in the coordinates of $v\in E$. We define $f(v)=(-1)^{N(v)}$ for $v\in E$ and f(v)=0 elsewhere. If $g=(a_1,a_2,\ldots,a_j)$ then the coset $H_ig=\{(a_1,a_2,\ldots,a_{i-1},h,a_{i+1},\ldots,a_j): h\in H_i\}$. If there exists k, $1\leq k\leq j$ $k\neq i$, such that $a_k\notin E_k$ then $H_ig\cap E=\emptyset$ and f is identically zero on H_ig. If $a_k\in E_k$ for every $k\neq i$, then

$$H_ig\cap E=\{(a_1,..,a_{i-1},x_i,a_{i+1},..,a_j),(a_1,..,a_{i-1},y_i,a_{i+1},..,a_j)\}.$$

Since

$$N((a_1,..,a_{i-1},x_i,a_{i+1},..,a_j))=N((a_1,..,a_{i-1},y_i,a_{i+1},..,a_j))+1,$$

it follows that $\sum_{h\in H_i} f(ht)=0$ for every t and $i=1,2,\ldots j$. This proves the lemma in the special case G=H. Finally, by Lemma (3.6) it is enough to find a function $F\not\equiv 0$ on G such that $\sum_{h\in H_i} F(ht)=0$ for every $t\in G$ and $i=1,2,\dots,j$. This is because the assumptions imply that every set gH_it is a left coset of H_k for some k (depending on g and i). Let f be the function defined above for the group $H=H_1\times H_2\times\dots\times H_j$, then $F=f\chi_H$ satisfies the requirements.∎

(3.8) *LEMMA. Let* $S(q+1)$ *be the group of all permutations of the set* $\{1,2,\dots,q+1\}$ $(q+1\geq 3)$. *Let* $S(q)$ *be the stability subgroup of a point of* $\{1,2,\dots,q+1\}$. *Then there exists a (unitary) irreducible representation of* $S(q+1)$ *which has no nontrivial* $S(q)$-*invariant vectors.*

PROOF. This is a consequence of a standard result in the theory of finite groups [I, Corollary 5.17] which asserts that if a finite group G with $|G|>2$ acts doubly transitively on a set X and H is the stability subgroup of a point of X then there exists an irreducible representation of G which is not contained in the quasi-regular representation on G/H. We include a direct proof for S(q+1) for completeness. First, we observe that if G is a finite nonabelian group and H is a subgroup of G such that every irreducible representation of G has a nontrivial H-invariant vector then $|G|/|H|>|\hat{G}|$ (where $\hat{G}$ is the dual object of G). This is a consequence of the fact that the irreducible representations of G having a nontrivial H-invariant vector are, up to equivalence, exactly the irreducible subrepresentations of the quasi-regular representation $\lambda_{G/H}$, that is the representation which corresponds to the left-invariant space consisting of H-right-invariant functions on G (see the proofs of Lemmas (2.4) and (2.5), or the Frobenius reciprocity theorem [I, Lemma 5.2]). Therefore $\lambda_{G/H}=\oplus_{\sigma\in\hat{G}} n_\sigma\sigma$ where $n_\sigma\geq 1$ iff σ has a nontrivial H-invariant vector. If $n_\sigma\geq 1$ for every σ then $|G|/|H|=\sum_{\sigma\in\hat{G}} n_\sigma \dim(\sigma)\geq$

$\sum_{\sigma\in\hat{G}}\dim(\sigma)\geq |\hat{G}|$. Since G is nonabelian there exists a σ such that $\dim(\sigma)>1$ and so $|G|/|H|>|\hat{G}|$. This proves the claim. Therefore to prove the lemma it suffices to show that $|S(q+1)|/|S(q)|=(q+1)!/q!=q+1\leq|\hat{S}(q+1)|$. This is obvious because $\hat{G}$ corresponds, bijectively, to the set of conjugacy classes of G; if $u_j\in S(q+1)$, $j=0,1,\ldots,q$, is a permutation which fixes exactly j points of $\{1,2,\ldots,q+1\}$ and C_j is the conjugacy class of u_j, then $C_i\cap C_j=\emptyset$ for $i\neq j$ and so $|\hat{S}(q+1)|\geq q+1$, and the lemma follows.∎

(3.9) *THEOREM. For every finite complete subtree* $\mathfrak{J}$ *with* $\operatorname{diam}(\mathfrak{J})\geq 2$ *the space* $(\tilde{K}(\mathfrak{J}))^{\wedge}_{o}\neq\emptyset$. *In particular* $\mathcal{S}(\mathfrak{J})\neq 0$.

PROOF. If $\operatorname{diam}(\mathfrak{J})=2$ then $\mathfrak{J}=V_1(\{x_o\})$ for some $x_o\in\mathfrak{X}$; $\tilde{K}(\mathfrak{J})/K(\mathfrak{J})\simeq \operatorname{Aut}(\mathfrak{J})\simeq S(q+1)$. The maximal proper complete subtrees of $\mathfrak{J}$ are the q+1 edges of $\mathfrak{J}$. $\operatorname{Aut}(\mathfrak{J})$ acts transitively on the edges and so the groups $K(\mathfrak{J}_i)/K(\mathfrak{J})\simeq S(q)$ are conjugate to each other. Since v is H-invariant iff $\pi(g)v$ is gHg^{-1}-invariant, by Lemma (3.8) $(\tilde{K}(\mathfrak{J}))^{\wedge}_{o}\neq\emptyset$. Let now $\mathfrak{J}$ be a complete finite subtree with $\operatorname{diam}(\mathfrak{J})>2$. Let $\mathfrak{J}^{o}$ be the subtree of $\mathfrak{J}$ consisting of vertices of homogeneity q+1. As in the proof of Lemma (3.1), $K(\mathfrak{J}^{o})/K(\mathfrak{J})$ decomposes into the direct product of its nontrivial subgroups $K(\mathfrak{J}_i)/K(\mathfrak{J})$ where $\mathfrak{J}_1,\mathfrak{J}_2,\ldots,\mathfrak{J}_j$ are the maximal proper complete subtrees of $\mathfrak{J}$. Since the inner automorphisms of $\operatorname{Aut}(\mathfrak{J})$ permute the subgroups $K(\mathfrak{J}_i)/K(\mathfrak{J})$, by Lemma (3.7) $(\tilde{K}(\mathfrak{J}))^{\wedge}_{o}\neq\emptyset$. Let $\sigma\in(\tilde{K}(\mathfrak{J}))^{\wedge}_{o}$ and $\xi\in\mathcal{H}_{\sigma}$ $\xi\neq 0$; then $(\sigma(.)\xi,\xi)$ is a $K(\mathfrak{J})$-bi-invariant nontrivial function on $\tilde{K}(\mathfrak{J})$ such that the right-averaging over $K(\mathfrak{J}_i)$ is zero for every i. Because $K(\mathfrak{J})\subset K(\mathfrak{J}_i)\subset\tilde{K}(\mathfrak{J})$ it follows that the function defined by $f(x)=(\sigma(x)\xi,\xi)$ for $x\in\tilde{K}(\mathfrak{J})$ and $f(x)=0$ elsewhere is in $\mathcal{S}(\mathfrak{J})$. Thus $\mathcal{S}(\mathfrak{J})\neq 0$ and the theorem is proved.∎

Let $\mathcal{M}(\mathfrak{J})$ be the subspace of $L^2(\operatorname{Aut}(\mathfrak{X}))$ consisting of

$K(\mathfrak{I})$-right-invariant functions such that the right-averaging over $K(\mathfrak{I}')$ is zero for every complete proper subtree $\mathfrak{I}'$ of $\mathfrak{I}$ (or, equivalently, for every maximal proper complete subtree $\mathfrak{I}'$ of $\mathfrak{I}$). We have that $\mathcal{S}(\mathfrak{I})\subset\mathcal{M}(\mathfrak{I})$ and $\mathcal{M}(\mathfrak{I})$ is a closed left-invariant nontrivial subspace of $L^2(\mathrm{Aut}(\mathfrak{X}))$. Let $\lambda_{\mathfrak{I}}$ be the subrepresentation obtained by restricting to $\mathcal{M}(\mathfrak{I})$ the left regular representation of $\mathrm{Aut}(\mathfrak{X})$. Obviously $P_\lambda(\mathfrak{I}')\mathcal{M}(\mathfrak{I})\subset\mathcal{S}(\mathfrak{I})$ for every $\mathfrak{I}'$.

(3.10) *PROPOSITION. Let* $\chi_{\mathfrak{I},\mathfrak{I}'}$ *be the characteristic function of the set* $\{g\in\mathrm{Aut}(\mathfrak{X}):\ g\mathfrak{I}\subset\mathfrak{I}'\}$. *If* $\mathfrak{I}$ *and* $\mathfrak{I}'$ *are complete finite subtrees of* $\mathfrak{X}$ *with* $\mathrm{diam}(\mathfrak{I})\geq 2$ *then*

(i) for every $f\in\mathcal{M}(\mathfrak{I})$, $P_\lambda(\mathfrak{I}')f=f\chi_{\mathfrak{I},\mathfrak{I}'}$;

(ii) $\mathcal{M}(\mathfrak{I})\cap C_c(\mathrm{Aut}(\mathfrak{X}))=\mathcal{S}(\mathfrak{I})$ *where* $C_c(\mathrm{Aut}(\mathfrak{X}))$ *is the space of continuous functions with compact support on* $\mathrm{Aut}(\mathfrak{X})$;

(iii) $\mathcal{S}(\mathfrak{I})$ *is a dense subspace of* $\mathcal{M}(\mathfrak{I})$.

PROOF. By Proposition (3.2), $\mathrm{supp}\, P_\lambda(\mathfrak{I}')f\subset\{g\in\mathrm{Aut}(\mathfrak{X}):\ g\mathfrak{I}\subset\mathfrak{I}'\}$. Therefore it suffices to prove that $P_\lambda(\mathfrak{I}')f(g)=f(g)$ if $g\mathfrak{I}\subset\mathfrak{I}'$. We have that

$$P_\lambda(\mathfrak{I}')f(g)= (m(K(\mathfrak{I}')))^{-1}\int_{K(\mathfrak{I}')} f(tg)dt$$

$$= (m(K(\mathfrak{I}')))^{-1}\int_{g^{-1}K(\mathfrak{I}')} f(gt)dt = (m(K(\mathfrak{I}')))^{-1}\int_{K(g^{-1}\mathfrak{I}')} f(gt)dt.$$

If $g\mathfrak{I}\subset\mathfrak{I}'$ then $\mathfrak{I}\subset g^{-1}\mathfrak{I}'$ and $K(g^{-1}\mathfrak{I}')\subset K(\mathfrak{I})$. Since f is $K(\mathfrak{I})$-right-invariant, f is constant on $gK(g^{-1}\mathfrak{I}')$ and $m(K(g^{-1}\mathfrak{I}'))=m(K(\mathfrak{I}'))$. It follows that $P_\lambda(\mathfrak{I}')f(g)=f(g)$. To prove (ii) it is enough to show that every function $f\in\mathcal{M}(\mathfrak{I})\cap C_c(\mathrm{Aut}(\mathfrak{X}))$ is $K(\mathfrak{I}')$-left-invariant for some complete finite subtree $\mathfrak{I}'$. Since $K(\mathfrak{I})$ is compact open there exist $g_1,g_2,\dots,g_n\in\mathrm{Aut}(\mathfrak{X})$ such that $\mathrm{supp}\, f\subset\bigcup_{i=1}^{n} g_iK(\mathfrak{I})$; this means that the set $\bigcup_{h\in\,\mathrm{supp}\, f} h\mathfrak{I}$ is finite because it is contained in $\bigcup_{i=1}^{n} g_i\mathfrak{J}$. In particular there exists a complete finite subtree $\mathfrak{J}'$ such that $\bigcup_{h\in \text{supp } f} h\mathfrak{J}\subset\mathfrak{J}'$. This implies that supp $f\subset\{g\in\text{Aut}(\mathfrak{X}):\ g\mathfrak{J}\subset\mathfrak{J}'\}$. By (i) $f=f\chi_{\mathfrak{J},\mathfrak{J}'}=P_\lambda(\mathfrak{J}')f$; this proves (ii).

Let x_o be a fixed vertex of $\mathfrak{X}$ and $\mathfrak{J}_n=V_n(\{x_o\})=\{y\in\mathfrak{X}: d(y,x_o)\le n\}$. Let $E_n=\{g\in\text{Aut}(\mathfrak{X}):\ g\mathfrak{J}\subset\mathfrak{J}_n\}$ and for $f\in\mathcal{M}(\mathfrak{J})$, let $f_n=f\chi_{E_n}$. By (i) $f_n=P_\lambda(\mathfrak{J}_n)f\in\mathcal{S}(\mathfrak{J})$. For every n, E_n is compact open and $E_n\subset E_{n+1}$, $\bigcup_{n=1}^{\infty} E_n=\text{Aut}(\mathfrak{X})$. This implies that $f_n\longrightarrow f$ in $L^2(\text{Aut}(\mathfrak{X}))$ and the proposition follows.∎

Proposition (3.10) implies that a function $f\in\mathcal{M}(\mathfrak{J})$ is $K(\mathfrak{J})$-bi-invariant iff $f=f\chi_{\tilde{K}(\mathfrak{J})}$. In particular the functions of $\mathcal{M}(\mathfrak{J})$ with compact support in $g\tilde{K}(\mathfrak{J})$ are exactly the functions $\{f\chi_{g\tilde{K}(\mathfrak{J})}:\ f\in\mathcal{M}(\mathfrak{J})\}=\lambda(g)(\{f\chi_{\tilde{K}(\mathfrak{J})}:\ f\in\mathcal{M}(\mathfrak{J})\})$ because $\mathcal{M}(\mathfrak{J})$ is left-invariant. This proves that every function in $\mathcal{S}(\mathfrak{J})$ is a finite sum of left translates of $K(\mathfrak{J})$-bi-invariant functions of $\mathcal{S}(\mathfrak{J})$ because supp $f\subset\bigcup_{i=1}^{n} g_i\tilde{K}(\mathfrak{J})$ for some $g_1,g_2,\dots,g_n$.

The following two results are proved exactly as the corresponding results were proved in Section 2 for special representations.

(3.11) *LEMMA. Every closed left-invariant nontrivial subspace of $\mathcal{M}(\mathfrak{J})$ contains a nontrivial* $K(\mathfrak{J})$*-bi-invariant function.*

PROOF. (See Lemma (2.4).) ∎

(3.12) *LEMMA. Every cuspidal irreducible unitary representation of* Aut($\mathfrak{X}$) *with minimal tree $\mathfrak{J}$ is unitarily equivalent to a subrepresentation of $\lambda_{\mathfrak{J}}$.*

PROOF. (See Lemma (2.5).) ∎

As in Section 2, Lemmas (3.11) and (3.12) and the fact that the subspace of $\mathscr{S}(\mathfrak{T})$ consisting of $K(\mathfrak{T})$-bi-invariant functions is finite-dimensional imply that $\mathfrak{T}$ is a minimal tree for each subrepresentation of $\lambda_{\mathfrak{T}}$ and that $\lambda_{\mathfrak{T}}$ is a finite sum of irreducible subrepresentations. Moreover, up to equivalence, the cuspidal irreducible representations of $\mathrm{Aut}(\mathfrak{X})$ with minimal tree $\mathfrak{T}$ are the inequivalent irreducible subrepresentation of $\lambda_{\mathfrak{T}}$. If $[\mathfrak{T}]\neq[\mathfrak{T}']$ then $\lambda_{\mathfrak{T}}$ and $\lambda_{\mathfrak{T}'}$ have no common components (see Corollary (3.4)), while $\lambda_{\mathfrak{T}}$ is unitarily equivalent to $\lambda_{g\mathfrak{T}}$ for every $g\in\mathrm{Aut}(\mathfrak{X})$. We prove now that the map

$$\sigma \longrightarrow \operatorname*{Ind}_{\tilde{K}(\mathfrak{T})\uparrow\mathrm{Aut}(\mathfrak{X})} \sigma$$

is a bijection from $(\tilde{K}(\mathfrak{T}))^{\wedge}_{o}$ onto the classes of inequivalent cuspidal irreducible representations of $\mathrm{Aut}(\mathfrak{X})$ with minimal tree $\mathfrak{T}$.

We recall, briefly, a few basic results concerning induced representations of groups in the special case of unimodular separable locally compact groups and compact open subgroups (for proofs and more details about induced representations we refer the reader to **[M]** and **[G]**). Let G be a unimodular separable locally compact group and K be a compact open subgroup of G. Let σ be a unitary representation of K; let $\mathfrak{H}^{\sigma}$ be the space of functions $f\colon G \longrightarrow \mathcal{H}_{\sigma}$ such that

(i) $f(xh)=\sigma(h^{-1})f(x)$ for every $x\in G$ and $h\in K$,

(ii) $\displaystyle\int_G \|f(g)\|^2 dg < +\infty.$

We have that $\displaystyle\int_G \|f(g)\|^2 dg= m(K) \sum_{x\in G/K} \|f(g)\|^2$ and $\mathfrak{H}^{\sigma}$ is a Hilbert space with the following scalar product

$$(f,g)=\int_G (f(x),g(x))dx \quad \text{for } f,g\in\mathfrak{H}^\sigma.$$

Observe that $\mathfrak{H}^\sigma$ is invariant under left translations. The unitary left regular representation of G on $\mathfrak{H}^\sigma$ is called the induced representation of G by σ and is denoted by $\mathrm{Ind}(\sigma)$. Hence $\mathrm{Ind}(\sigma)(x)f(t)=f(x^{-1}t)$. If σ_1 is unitarily equivalent to σ_2 then also $\mathrm{Ind}(\sigma_1)$ is unitarily equivalent to $\mathrm{Ind}(\sigma_2)$ and $\mathrm{Ind}(\sigma_1\oplus\sigma_2)\simeq \mathrm{Ind}(\sigma_1)\oplus\mathrm{Ind}(\sigma_2)$. Let $\xi\in\mathcal{H}_\sigma$; we define $f_\xi(x)=\sigma(x^{-1})\xi$ for $x\in K$ and $f_\xi(x)=0$ for $x\notin K$. It follows that $f_\xi\in\mathfrak{H}^\sigma$ and $\|f\|=\|\xi\|\sqrt{m(K)}$. It is easy to see that $(\mathrm{Ind}(\sigma)(x)f_\xi,f_\xi)=(\sigma(x)\xi,\xi)(m(K))$ for $x\in K$ and $(\mathrm{Ind}(\sigma)(x)f_\xi,f_\xi)=0$ for $x\notin K$. This proves that every coefficient of σ is a coefficient of $\mathrm{Ind}(\sigma)$ with support in K. In particular this proves that if σ_1 and σ_2 are inequivalent then also $\mathrm{Ind}(\sigma_1)$ and $\mathrm{Ind}(\sigma_2)$ are inequivalent. If $\mathrm{Ind}(\sigma)$ is irreducible then it is an L^1-representation and it has a dense subset of coefficients with compact support; its formal dimension is equal to $\dim(\sigma)/m(K)$. In fact if $\xi\in\mathcal{H}_\sigma$, $\|\xi\|=1$ then

$$d_{\mathrm{ind}\sigma}=\left(\int_K |(\sigma(g)\xi,\xi)|^2dg\right)^{-1}=\left[m(K)\int_K |(\sigma(g)\xi,\xi)|^2(dg/m(K))\right]^{-1}$$

$$= d_\sigma/m(K) = \dim(\sigma)/m(K).$$

Let $\mathfrak{H}^\sigma_K$ be the subspace of $\mathfrak{H}^\sigma$ consisting of function with support in K. It is clear that $\mathfrak{H}^\sigma_K$ is a closed nontrivial subspace of $\mathfrak{H}^\sigma$; the map $\xi\to f_\xi$ is an isomorphism of $\mathcal{H}_\sigma$ onto $\mathfrak{H}^\sigma_K$. The map $f\to f\chi_K$ is the projection of $\mathfrak{H}^\sigma$ onto $\mathfrak{H}^\sigma_K$. This implies that every function of $\mathfrak{H}^\sigma$ with compact support is a linear combination of left translates of functions in $\mathfrak{H}^\sigma_K$ (see the remarks after Proposition (3.10)). In particular the subspace of $\mathfrak{H}^\sigma$ generated by $\bigcup_{g\in G}\lambda(g)\mathfrak{H}^\sigma_K$ is dense in $\mathfrak{H}^\sigma$. For completeness we include here the proof of the following proposition.

(3.13) *PROPOSITION. Let* σ *be a unitary irreducible representation of* K*; then* $\mathrm{Ind}(\sigma)$ *is irreducible if (and only if) every closed nontrivial invariant subspace of* $\mathfrak{H}^{\sigma}$ *contains a nontrivial function of* $\mathfrak{H}^{\sigma}_{K}$.

PROOF. It is enough to prove that if M is a closed nontrivial invariant subspace of $\mathfrak{H}^{\sigma}$ such that $\mathfrak{H}^{\sigma}_{K}\cap M\neq 0$ then $M=\mathfrak{H}^{\sigma}$. Since M is closed and $\mathrm{Ind}(\sigma)$-invariant it suffices to prove that $\mathfrak{H}^{\sigma}_{K}\subset M$ because, as observed, the subspace generated by $\bigcup_{g\in G}\mathrm{Ind}(\sigma)(g)\mathfrak{H}^{\sigma}_{K}$ is dense in $\mathfrak{H}^{\sigma}$. Let $N=\{f(1_G):\ f\in M\cap\mathfrak{H}^{\sigma}_{K}\}$. We recall that if $f\in\mathfrak{H}^{\sigma}_{K}$ then $f\equiv 0$ iff $f(1_G)=0$; so $M\cap\mathfrak{H}^{\sigma}_{K}\neq 0$ implies that $N\neq 0$. Moreover, for every $k\in K$ we have that $\sigma(k)f(1_G)=f(k^{-1})=(\mathrm{Ind}(\sigma)(k)f)(1_G)$ and $\mathrm{Ind}(\sigma)(k)f\in\mathfrak{H}^{\sigma}_{K}\cap M$; this means that $\sigma(k)f(1_G)\in N$ and N is a nontrivial σ-invariant subspace of $\mathcal{H}_{\sigma}$. Since σ is irreducible it follows that $N=\mathcal{H}_{\sigma}$. Let $g\in\mathfrak{H}^{\sigma}_{K}$; then there exists $f\in M\cap\mathfrak{H}^{\sigma}_{K}$ such that $f(1_G)=g(1_G)$; therefore $f(k)=g(k)$ for every $k\in K$ and so $f\equiv g$. This means that $\mathfrak{H}^{\sigma}_{K}\subset M$ and the proposition follows.∎

(3.14) *THEOREM. Let* $\mathfrak{J}$ *be a finite complete subtree of* $\mathfrak{X}$ *with* $\mathrm{diam}(\mathfrak{J})\geq 2$. *Let* $\sigma_{\mathfrak{J}}=\bigoplus_{\sigma\in(\tilde{K}(\mathfrak{J}))^{\wedge}_{0}}\dim(\sigma)\sigma$*; then* $\lambda_{\mathfrak{J}}$ *is unitarily equivalent to* $\mathrm{Ind}(\sigma_{\mathfrak{J}})=\bigoplus_{\sigma\in(\tilde{K}(\mathfrak{J}))^{\wedge}_{0}}\dim(\sigma)\mathrm{Ind}(\sigma)$. *For every* $\sigma\in(\tilde{K}(\mathfrak{J}))^{\wedge}_{0}$, $\mathrm{Ind}(\sigma)$ *is irreducible and the map* $\sigma\longrightarrow\mathrm{Ind}(\sigma)$ *is a bijection from* $(\tilde{K}(\mathfrak{J}))^{\wedge}_{0}$ *onto the inequivalent irreducible subrepresentations of* $\lambda_{\mathfrak{J}}$ *, that is the classes of distinct irreducible cuspidal representations of* $\mathrm{Aut}(\mathfrak{X})$ *with minimal tree* $\mathfrak{J}$. *The formal dimension of* $\mathrm{Ind}(\sigma)$ *is equal to* $\dim(\sigma)/m(\tilde{K}(\mathfrak{J}))$.

PROOF. Let M be the subspace of $\mathcal{S}(\mathfrak{J})$ consisting of $K(\mathfrak{J})$-bi-invariant functions. Then M is a nontrivial $\tilde{K}(\mathfrak{J})$-bi-invariant subspace of functions with support in $\tilde{K}(\mathfrak{J})$ (by (3.2)). Let $\sigma_{\mathfrak{J}}$ be the subrepresentation of $\lambda_{\tilde{K}(\mathfrak{J})}$ relating to M. Then M is finite-dimensional and so $\sigma_{\mathfrak{J}}$ is a finite sum of irreducible representations. Finally, $\sigma_{\mathfrak{J}}$ is standard and it is trivial on $K(\mathfrak{J})$, so every irreducible subrepresentation of $\sigma_{\mathfrak{J}}$ is in $(\tilde{K}(\mathfrak{J}))^{\wedge}_{o}$. Conversely, if σ is a standard subrepresentation of $\lambda_{\tilde{K}(\mathfrak{J})}$, trivial on $K(\mathfrak{J})$, then the left-invariant subspace of $L^2(\tilde{K}(\mathfrak{J}))$ which corresponds to σ is contained in M and so σ is contained in $\sigma_{\mathfrak{J}}$; in particular

$$\bigoplus_{\sigma\in(\tilde{K}(\mathfrak{J}))^{\wedge}_{o}} (\dim(\sigma))\sigma<\sigma_{\mathfrak{J}}.$$

Since

$$\lambda_{\tilde{K}(\mathfrak{J})} = \bigoplus_{\pi\in(\tilde{K}(\mathfrak{J}))^{\wedge}_{o}} (\dim(\pi))\pi$$

it follows that $\sigma_{\mathfrak{J}} = \bigoplus_{\sigma\in(\tilde{K}(\mathfrak{J}))^{\wedge}_{o}} (\dim(\sigma))\sigma$. We prove now that $\lambda_{\mathfrak{J}}$ is unitarily equivalent to $\mathrm{Ind}(\sigma_{\mathfrak{J}})$. Let $\mathfrak{H}^{\sigma_{\mathfrak{J}}}_{x\tilde{K}(\mathfrak{J})}$ be the closed subspace of $\mathfrak{H}^{\sigma_{\mathfrak{J}}}$ consisting of functions with support in $x\tilde{K}(\mathfrak{J})$. We put $\mathfrak{U}f=\lambda(x)[f(x)]$ for $f\in\mathfrak{H}^{\sigma_{\mathfrak{J}}}_{x\tilde{K}(\mathfrak{J})}$ (we recall that $f:\mathrm{Aut}(\mathfrak{X})\longrightarrow M$). In particular, $\mathfrak{U}f$ is a function in $\mathcal{S}(\mathfrak{J})$ which is $K(x\mathfrak{J})$-left-invariant. This operator is well defined, in fact $\lambda(x)[f(x)]=\lambda(xk)[f(xk)]$ if $k\notin\tilde{K}(\mathfrak{J})$. $\mathfrak{U}$ is a linear bijective operator from $\mathfrak{H}^{\sigma_{\mathfrak{J}}}_{x\tilde{K}(\mathfrak{J})}$ onto the subspace of $K(x\mathfrak{J})$-left-invariant functions of $\mathcal{S}(\mathfrak{J})$ which is equal to $\lambda(x)M$. In fact if $\xi\in M$ the function $f(xk)=\lambda(k^{-1})\xi$ for $k\in\tilde{K}(\mathfrak{J})$ and $f=0$ elsewhere is a function in $\mathfrak{H}^{\sigma_{\mathfrak{J}}}_{x\tilde{K}(\mathfrak{J})}$ such that $\mathfrak{U}f=\lambda(x)\xi$, which proves that $\mathfrak{U}$ is surjective. The functions in $\mathfrak{H}^{\sigma_{\mathfrak{J}}}_{x\tilde{K}(\mathfrak{J})}$ depend only on the value in x, so $\mathfrak{U}f=\mathfrak{U}g$ implies that $f(x)=g(x)$ and $\mathfrak{U}$ is injective. We choose the Haar measure of $\mathrm{Aut}(\mathfrak{X})$ in such a way that $\tilde{K}(\mathfrak{J})$ has measure equal to 1. In this way $\mathfrak{U}$ is isometric. We recall that every function of $\mathcal{S}(\mathfrak{z})$ is a linear combination of $K(x\mathfrak{z})$-left-invariant functions and $\mathcal{S}(\mathfrak{z})$ is dense in $\mathcal{M}(\mathfrak{z})$. On the other hand the space of functions of $\mathfrak{H}^{\sigma_{\mathfrak{z}}}$ with compact support is dense in $\mathfrak{H}^{\sigma_{\mathfrak{z}}}$ and every such function is a finite sum of functions in $\mathfrak{H}^{\sigma_{\mathfrak{z}}}_{x\tilde{K}(\mathfrak{z})}$. If $x\tilde{K}(\mathfrak{z})\neq y\tilde{K}(\mathfrak{z})$ then $\mathfrak{H}^{\sigma_{\mathfrak{z}}}_{x\tilde{K}(\mathfrak{z})} \cap \mathfrak{H}^{\sigma_{\mathfrak{z}}}_{y\tilde{K}(\mathfrak{z})} = \{0\}$ and so $\mathfrak{U}$ extends to a unitary operator from $\mathfrak{H}^{\sigma_{\mathfrak{z}}}$ onto $\mathcal{M}(\mathfrak{z})$ intertwining $\mathrm{Ind}(\sigma_{\mathfrak{z}})$ and $\lambda_{\mathfrak{z}}$. Under the action of $\mathfrak{U}$, the condition of Proposition (3.13) is exactly Lemma (3.11), and hence $\mathrm{Ind}(\sigma)$ is irreducible for every $\sigma\in(\tilde{K}(\mathfrak{z}))^{\wedge}_{o}$. The rest of the statement follows from the previous remarks. ∎

Putting together the results of Chapter II and Section 2 for $G=\mathrm{Aut}(\mathfrak{X})$, we have a complete classification of all unitary irreducible representations of $\mathrm{Aut}(\mathfrak{X})$.

We observe now that $\mathrm{Aut}(\mathfrak{X})$ is a type I group. A unitary representation π is called admissible if the spaces $\mathcal{H}_{\pi}(\mathfrak{z})$ are finite-dimensional for every $\mathfrak{z}$. We prove that all unitary irreducible representations of $\mathrm{Aut}(\mathfrak{X})$ are admissible. Indeed, if $\mathfrak{z}$ is a minimal tree of π and ξ is a nontrivial $K(\mathfrak{z})$-invariant vector, then the map $\eta\longrightarrow\overline{(\pi(.)\xi,\eta)}$ is a linear injective map from $\mathcal{H}_{\pi}$ in the space of $K(\mathfrak{z})$-right-invariant functions, as is easily seen because π is irreducible and ξ is a cyclic vector. In particular if π is a special or cuspidal representation then, for every $\mathfrak{z}'$, $\mathcal{H}_{\pi}(\mathfrak{z}')$ is isomorphic to a subspace of the space of $K(\mathfrak{z}')$-left-invariant functions of $\mathcal{S}(\mathfrak{z})$ which is finite-dimensional (remarks after Lemma (2.1) and Corollary (3.4)). The proof of the fact that every spherical irreducible representation of $\mathrm{Aut}(\mathfrak{X})$ is admissible is similar. Indeed let $\mathfrak{z}\subseteq\mathfrak{X}_{n}=\{x\in\mathfrak{X}\colon\ d(x,o)\leq n\}$ and, with the terminology and notation of (II,5), let $V=\{\xi\in\mathcal{K}(\Omega): \pi_z(k)\xi=\xi$ for every $k\in K(\mathfrak{X}_n)\}$. Then V has dimension $(q+1)q^{n-1}$. Therefore π_z is admissible. Since every spherical representation is equivalent to π_z for some $z\in\mathbb{C}$, such that $-1\le\mu(z)\le1$, we conclude that every spherical representation is admissible.

These facts imply that $\pi(L^1(\mathrm{Aut}(\mathfrak{X})))$ contains a nontrivial compact operator, indeed if $\mathfrak{I}$ is a minimal tree of π then $0\ne P_\pi(\mathfrak{I})=\pi(f)$ for $f=(m(K(\mathfrak{I})))^{-1}\chi_{K(\mathfrak{I})}$. By **[D1, 13.9.4]** Aut($\mathfrak{X}$) is *postliminal* and so Aut($\mathfrak{X}$) is a type I group **[D1**, 18.8.2].

Finally we remark that, in the terminology of **[D1**, 13.9.4], Aut($\mathfrak{X}$) is *liminal*, that is $\pi(f)$ is a compact operator for every $f\in L^1(\mathrm{Aut}(\mathfrak{X}))$. This is a consequence of the facts that the linear space F generated by the characteristic functions of the sets $gK(\mathfrak{I}')$, for every $g\in\mathrm{Aut}(\mathfrak{X})$ and every complete subtree $\mathfrak{I}'$, is dense in $L^1(\mathrm{Aut}(\mathfrak{X}))$, and $\pi(f)$ is a compact operator for every $f\in F$ because π is admissible. In particular the closure of $\pi(L^1(\mathrm{Aut}(\mathfrak{X})))$ in the norm operator topology is $\mathcal{LC}(\mathcal{H}_\pi)$ the space of all compact operators on $\mathcal{H}_\pi$.

Every irreducible representation of Aut($\mathfrak{X}$) is either spherical or square-integrable; therefore the Plancherel spherical formula (II,(6.5)) and the expressions for the formal dimensions of special and cuspidal irreducible representations of Aut($\mathfrak{X}$) yield the Plancherel formula of Aut($\mathfrak{X}$) **[D1**;18.8.2,18.8.5]. Indeed, let $[\mathfrak{I}]=\{g\mathfrak{I}: g\in\mathrm{Aut}(\mathfrak{X})\}$; let π^+, π^- be the special representations of Aut($\mathfrak{X}$). We choose the Haar measure of Aut($\mathfrak{X}$) in such a way that the stability subgroup of a vertex has measure equal to 1. Therefore for every continuous function f on Aut($\mathfrak{X}$) with compact support we have

$$\|f\|_2^2 = \int_J \|\pi_{1/2+it}(f)1\|^2_{L^2(\Omega)} dm(t)$$

$$+[(q-1)/2][tr(\pi^+(f)\pi^+(f)^*) + tr(\pi^-(f)\pi^-(f)^*]$$

$$+ \sum_{\substack{[\mathfrak{z}] \\ diam(\mathfrak{z})\geq 2}} (m(\tilde{K}(\mathfrak{z})))^{-1}[\sum_{\sigma\in(\tilde{K}(\mathfrak{z}))^\wedge_o} dim(\sigma)\ tr(Ind\sigma(f)Ind\sigma(f)^*)].$$

4. Notes and Remarks. The content of the present Chapter is basically due to G.I. Ol'shianskii and is taken from **[O2]**; in particular Proposition (3.2), which is the crucial step for the classification of all cuspidal irreducible representations of Aut($\mathfrak{X}$), is due to G.I. Ol'shianskii who gives in **[O2]** the classification of algebraic, admissible and algebraically irreducible representations of Aut($\mathfrak{X}$) in a vector space V (*a priori* with no topology). Our exposition however is more direct insofar as it does not make use of the notion of admissible representation. We have chosen also to treat special representations for a group more general than Aut($\mathfrak{X}$). Our treatment applies for instance to **PGL**(2,$\mathfrak{F}$).

We now give a brief description of Ol'shianskii's original approach and the connection between the present chapter and **[O2]**.

A representation π in V is called *algebraic* if each vector v is K($\mathfrak{z}$)-invariant for some $\mathfrak{z}$ depending on v. An *admissible* representation π is such that the subspace of K($\mathfrak{z}$)-invariant vectors is finite-dimensional for every fixed complete finite subtree $\mathfrak{I}$. Finally, π is said to be *algebraically irreducible* if there are no nontrivial invariant subspaces of V. We have observed that if π is a unitary irreducible representation of Aut($\mathfrak{X}$) in a Hilbert space $\mathcal{H}_\pi$ then V_π is a dense nontrivial invariant subspace of $\mathcal{H}_\pi$ (see the remark after Definition (1.1)) and π is admissible. On the other hand by **[O2]** it follows that every algebraic, admissible, algebraically irreducible representation of Aut($\mathfrak{X}$) is unitary. Let now π be a unitary irreducible representation in $\mathcal{H}_\pi$; let π^o be the restriction of π to the dense invariant subspace V_π. We have that π^o is an algebraic, admissible, algebraically irreducible representation of Aut($\mathfrak{X}$) (in fact if M is a nontrivial invariant subspace of V_π then M is dense in $\mathcal{H}_\pi$ because π is irreducible; therefore $P_\pi(\mathfrak{I})M$ is dense in $\mathcal{H}_\pi(\mathfrak{I})=P_\pi(\mathfrak{I})\mathcal{H}_\pi$; this means that $P_\pi(\mathfrak{I})M=\mathcal{H}_\pi(\mathfrak{I})$ for every $\mathfrak{I}$ because $\mathcal{H}_\pi(\mathfrak{I})$ is finite-dimensional. Since for every $\mathfrak{I}$ and $v\in V_\pi$, $P_\pi(\mathfrak{I})v$ is a finite linear combination of $\pi(k)v$ with $k\in K(\mathfrak{I})$, it follows that $P_\pi(\mathfrak{I})M\subset M$ for every $\mathfrak{I}$ and so $V_\pi=M$). Also if π_1 and π_2 are two unitary irreducible representations then π_1 is unitarily equivalent to π_2 if and only if π_1^o is algebraically equivalent to π_2^o, that is iff there exists a linear isomorphism T of V_{π_1} onto V_{π_2} such that $T\pi_1(g)=\pi_2(g)T$ for every $g\in$Aut($\mathfrak{X}$). In fact if such a T exists, then $T(\mathcal{H}_{\pi_1}(\mathfrak{I}))=\mathcal{H}_{\pi_2}(\mathfrak{I})$ and $TP_{\pi_1}(\mathfrak{I})=P_{\pi_2}(\mathfrak{I})T$ for every finite complete subtree $\mathfrak{I}$ (we recall that if v is $K(\mathfrak{I}')$-invariant for some $\mathfrak{I}'$ then, without loss of generality, we can suppose that $\mathfrak{I}\subset\mathfrak{I}'$ and so $K(\mathfrak{I}')$ has finite index in $K(\mathfrak{I})$, therefore $P_\pi(\mathfrak{I})v$ is a finite linear combination of $\pi(k)v$ with $k\in K(\mathfrak{I})$ and $TP_{\pi_1}(\mathfrak{I})v=P_{\pi_2}(\mathfrak{I})Tv$). This proves that if $v_n\in V_{\pi_1}$, $v_n\to 0$ and $Tv_n\to w$ then $w=0$ (in fact $P_{\pi_2}(\mathfrak{I})Tv_n=TP_{\pi_1}(\mathfrak{I})v_n\to 0$ and so $P_{\pi_2}(\mathfrak{I})w=0$ for every $\mathfrak{I}$). By [**Y**, Prop.2, p.77] there exists a closed operator $\overline{T}$ defined on a domain D which includes V_{π_1} and which agrees with T on V_{π_1}. In addition $\overline{T}\pi_1(g)=\pi_2(g)\overline{T}$ on D. The operator $\overline{T}$ may be uniquely written as the product $\overline{T}$=US where U is a partial isometry and $S=\sqrt{\overline{T}^* \overline{T}}$. The irreducibility of π_1 and π_2 implies that U is a unitary operator intertwining π_1 and π_2 (see also [JL, Lemma 2.6, p.30]).

It follows that the classes (up to algebraic equivalence) of algebraic, admissible, algebraically irreducible representations of Aut($\mathfrak{X}$) in a vector space are exactly the representations π^o where π varies among all classes (up to unitary equivalence) of unitary (topologically) irreducible representations of Aut($\mathfrak{X}$) in a Hilbert space. For instance, the invariant irreducible subspaces of $\mathcal{M}(\mathfrak{J})$ are the closures in L^2(Aut($\mathfrak{X}$)) of the algebraically irreducible invariant subspaces of $\mathcal{S}(\mathfrak{J})$.

A similar result is true for special representations of a closed subgroup acting transitively on $\mathfrak{X}$ and on Ω.

We recall that G.I.Ol'shanskii in **[O1]** proves that Aut($\mathfrak{X}$) is a type I group.

Observe that Ol'shianskii's method does not apply effectively to the classification of irreducible representations of other notable subgroups of Aut($\mathfrak{X}$). As noticed at the beginning of Section 3, Proposition (3.2) does not apply to **PGL**(2,$\mathfrak{F}$) because it does not contain sufficiently many rotations. Thus while the notion of *"cuspidal"* representation may be formally considered for any closed subgroup of Aut($\mathfrak{X}$), when the group considered does not contain sufficiently many rotations, it may happen that *"cuspidal representations"* fail to have coefficients with compact support. This is exactly what happens for **PGL**(2,$\mathfrak{F}$) since there exist nonspherical irreducible representations which are not square-integrable (the nonspherical principal series). The classification of irreducible representations of **PGL**$(2,\mathfrak{F})$ may be achieved using the machinery of **[GGP]** or **[Si]**. Of course spherical and special unitary representations of **PGL**$(2,\mathfrak{F})$ are exactly the restrictions of the corresponding representations of Aut$(\mathfrak{X})$. But the representations of the nonspherical principal series cannot be obtained as restrictions of irreducible representations of Aut$(\mathfrak{X})$.

On the other hand Ol'shianskii's ideas may be applied to classify the irreducible representations of the group B_ω which fixes a point ω of the boundary and stabilizes the horocycles **[N4]**. Observe that $B_\omega \cap$**PGL**$(2,\mathfrak{F})$ with an appropriate choice of ω is the group of matrices $\begin{bmatrix} a & b \\ 0 & a^{-1} \end{bmatrix}$ with $a,b \in \mathfrak{F}$ and $|a|=1$ (see Appendix Proposition (5.6), below).

We remark that cuspidal and special representations, unlike spherical representations, do not restrict irreducibly to any discrete subgroup of Aut$(\mathfrak{X})$; the reason being that no irreducible representations of an infinite discrete group may be contained in the regular representation **[C F-T]**. A.M.Mantero and A.Zappa **[MZ2]** also study the same restriction to a free group proving a very weak form of *irreducibility*. The decomposition of the restriction of special representations to a simply transitive subgroup of a tree was analyzed in full by G.Kuhn and T.Steger **[KS1]**.

We summarize the results on irreducibility of restrictions of representations of Aut$(\mathfrak{X})$ as follows.

(1) Spherical representations restrict irreducibly to any closed unimodular subgroup acting transitively on $\mathfrak{X}$ **[B-K1]** and to lattices (except for the exceptions indicated in Theorem (II,7.1) **[CS]** **[St2]**.

(2) Special representations restrict irreducibly to a closed subgroup acting transitively on $\mathfrak{X}$ and Ω (see Section 2).

(3) Cuspidal representations do not restrict irreducibly,

in general, to a group acting transitively on $\mathfrak{X}$ and Ω. We know for instance that some cuspidal representations must reduce when restricted to **PGL**$(2,\mathfrak{F})$. Indeed every coefficient of the regular representation of **PGL**$(2,\mathfrak{F})$ is the restriction of a coefficient of the regular representation of Aut$(\mathfrak{X})$ **[Hz1]** **[Hz2]**. On the other hand one can consider a subrepresentation of the regular representation of **PGL**$(2,\mathfrak{F})$ which is a direct integral of representations of the nonspherical principal series. This representation is orthogonal to the spherical representations and to the special representations and therefore can only be obtained by restricting cuspidal representations.

Unitary representations of the group of automorphisms of a homogeneous (or semihomogeneous) tree which is not locally finite have also been studied by Ol'shianskii **[O3]**. This group is not locally compact but turns out to be of type I.

The full Plancherel formula for Aut$(\mathfrak{X})$ is given here for the first time as a consequence of Olshianskii's work and the spherical Plancherel formula. The Plancherel formula for **SL**$(2,\mathfrak{F})$ was given by P.J. Sally and J.A. Shalika **[SS]**. C.L. Gulizia **[Gu]** made use of the full Plancherel formula to prove the Kunze-Stein phenomenon for **SL**$(2,\mathfrak{F})$ where $\mathfrak{F}$ is a local field such that the finite field $\mathfrak{O}/\mathfrak{P}$ has characteristic $p\neq 2$ (see Appendix below). Notice however that only the spherical Plancherel formula is needed for the Kunze-Stein phenomenon as shown in **[N3]** where the Kunze-Stein phenomenon is established for any closed noncompact subgroup acting transitively on Ω and therefore also for **PSL**$(2,\mathfrak{F})$ without restrictions on the characteristic of $\mathfrak{O}/\mathfrak{P}$.

A curious fact is that the Kunze-Stein phenomenon is equivalent to transitive action on Ω when $q+1=3$ **[N4]**.

We thank J. McMullen who made available to us his notes **[MM]** on Ol'shianskii's work.

APPENDIX

1. p-adic fields. We will define in this section a class of locally compact fields which can obtained by completing the field of rational numbers with respect to a metric different from the ordinary metric. Let $\mathbb{Q}$ be the field of rational numbers and $\mathbb{Q}^*$ the set of nonzero elements. Let p be a fixed prime. If $x\in\mathbb{Q}^*$ then $x=p^h m/n$, where $h,m,n\in\mathbb{Z}$, $n>0$ and $(m,n)=(n,p)=(m,p)=1$.
In other words m and n have no common divisors and neither of them is divisible by p. Such a triple h, m, n is uniquely identified by x, and therefore the map $x\longrightarrow|x|_p=p^{-h}$ is well defined for $x\in\mathbb{Q}^*$. We complete the definition by letting $|0|_p=0$. The number $|x|_p$ is sometimes called the p-adic norm or modulus of the rational number x. It is not difficult to show that the norm satisfies the following properties:

$$(i)\qquad |xy|_p = |x|_p|y|_p \;;$$

$$(ii)\qquad |x + y|_p \le \max(|x|_p, |y|_p).$$

Property *(ii)* is called the *ultrametric inequality* . It obviously implies that

$$(iii)\qquad |x + y|_p \le |x|_p + |y|_p.$$

It follows that the norm induces a metric $d(x,y)=|x-y|_p$, with respect to which the field operations in $\mathbb{Q}$ are continuous. As a consequence the completion of $\mathbb{Q}$ with respect to this metric is a field which is denoted by $\mathbb{Q}_p$. The sets $\{x\in\mathbb{Q}_p: |x|_p<p^n\}$ for $n\in\mathbb{Z}$ are a basis of a neighborhood of zero. Notice that the integers $\mathbb{Z}\subset\mathbb{Q}_p$ have norm less than or equal to 1, and that the integers not divisible by p have norm 1. The closure of the integers in $\mathbb{Q}_p$ is denoted by $\mathbb{Z}_p$. We will show

now that $\mathbb{Z}_p$ is *compact* and that $\mathbb{Z}_p=\{x: |x|_p\le 1\}$. To prove the first assertion it is sufficient to show that $\mathbb{Z}$ is totally bounded in $\mathbb{Q}_p$. Indeed, let $\varepsilon>0$ and $p^{-n}<\varepsilon$. For $k=0,1,2,\dots,p-1$, let

$$S(\varepsilon,k) = \{x\in\mathbb{Q}_p: |x-k|_p<\varepsilon\}.$$

If $h\in\mathbb{Z}$, then, for some $k=0,1,2,\dots,p-1$, $h-k\equiv 0 \bmod p^n$. Therefore $h\in S(\varepsilon,k)$. This shows that $\mathbb{Z}\subset\bigcup_{k=0}^{p-1} S(\varepsilon,k)$ and that $\mathbb{Z}$ is totally bounded. To show that $\mathbb{Z}_p=\{x: |x|_p\le 1\}$, let $|x|_p\le 1$ $x\in\mathbb{Q}$; then $x=p^r m/n$ with $(n,m)=(n,p)=(m,p)=1$, $n>0$ and $r\ge 0$. We prove that given $\varepsilon>0$ there exists an integer at distance less than ε from x. Let $p^{-k}<\varepsilon$ and observe that $(n,p)=1$ implies that n is invertible modulo p^k. Therefore there exists an integer $h\in\mathbb{Z}$ such that $hn\equiv 1 \bmod p^k$. Thus $|x-p^r hm|_p\le p^{-r-k}<\varepsilon$. This means that $\mathbb{Z}_p=\{x: |x|_p\le 1\}$. Observe that $\{x: |x|_p\le p^n\}=\{x: |x|_p<p^{n+1}\}$, because the norm only attains the values p^n, $n\in\mathbb{Z}$. Therefore the base for the neighborhood system of zero consists of compact open subsets. In conclusion $\mathbb{Q}_p$ is a *totally disconnected locally compact field of characteristic zero.*

2. A locally compact field of characteristic p. Let S_p be the ring of formal power series $\sum_{n\ge s} a_n X^n$, with $a_n\in\mathbb{Z}(p)$, the prime field of order p. We denote by $\tilde{a}_n$ the integer $0\le\tilde{a}_n<p$ representing the class $a_n\in\mathbb{Z}(p)$. With this notation the map $\sum_{n\ge s} a_n X^n \longrightarrow \sum_{n\ge s}\tilde{a}_n p^n$ is well defined with domain S_p and range contained in $\mathbb{Q}_p$. Indeed the series $\sum_{n\ge s}\tilde{a}_n p^n$ converges in $\mathbb{Q}_p$. Furthermore every element of $\mathbb{Q}_p$ can be expressed as a series $x=\sum_{n\ge s}\tilde{a}_n p^n$. For the proof of this fact we may assume that $|x|_p=1$. Then for $k\in\{0,1,2,\dots,p-1\}$ the spheres $S(p^{-1},k)=\{x\in\mathbb{Z}_p: |x-k|_p\le p^{-1}\}$ are disjoint and cover $\mathbb{Z}_p$. Therefore we can choose uniquely $\tilde{a}_0\in\{0,1,2,\dots,p-1\}$, such that $|x-\tilde{a}_0|_p\le p^{-1}$. Suppose by induction hypothesis that numbers $\tilde{a}_0,\tilde{a}_1,\dots,\tilde{a}_n\in\{0,1,2,\dots,p-1\}$ have been found to satisfy $|x-\sum_{j=0}^{n}\tilde{a}_j p^j|_p<p^{-n}$;

then

$$x-\sum_{j=0}^{n}\tilde{a}_j p^j = p^{n+1}y, \text{ with } |y|_p \le 1.$$

But y belongs to one and only one of the spheres $S(k,p^{-1})$, in other words there exists $\tilde{a}_{n+1}\in\{0,\dots,p-1\}$ such that $|y-\tilde{a}_{n+1}|_p<p^{-1}$. It follows that

$$|x-\sum_{j=0}^{n+1}\tilde{a}_j p^j|_p = |yp^{n+1}-\tilde{a}_{n+1}p^n|_p < p^{-n-2}.$$

In conclusion the map $\sum_{n\ge s} a_n X^n \longrightarrow \sum_{n\ge s}\tilde{a}_n p^n$ is surjective. We have also proved that it is injective because the construction above gives a unique sequence $\tilde{a}_n\in\{0,1,2,\dots,p-1\}$ and each $\tilde{a}_n$ identifies only one element of the residue classes modulo p. A formal power series can be given the norm $|\sum_{n\ge s} a_n X^n|_p = |\sum_{n\ge s}\tilde{a}_n p^n|_p = p^{-s}$. The map defined above becomes in this fashion an isometry and Sp is a complete metric space with respect to the distance $|x-y|_p$. It is important to keep in mind that the map $\sum_{n\ge s} a_n X^n \longrightarrow \sum_{n\ge s}\tilde{a}_n p^n$ is not a homomorphism. The operations of sum and product in Sp are those of the ring of power series with coefficients in $\mathbb{Z}(p)$. In particular Sp has characteristic p. On the contrary the operations on the *convergent* series $\sum_{n\ge s}\tilde{a}_n p^n$ are performed considering the elements $\tilde{a}_n$ as ordinary integers. Nevertheless the norm $|\cdot|_p$ in Sp satisfies the ultrametric inequality $|x+y|_p \le \max(|x|_p, |y|_p)$, as is immediately clear from the definitions. The multiplicative property $|xy|_p = |x|_p|y|_p$ is also evident. This means that Sp is a locally compact ring, which is complete in the metric defined by $|\cdot|_p$. We will presently show that Sp is a field. Observe first that elements of the type $a_n X^n$, with $n\in\mathbb{Z}$, are invertible, the inverse being simply $a_n^{-1}X^{-n}$, where $a_n^{-1}a_n=1$ in $\mathbb{Z}(p)$. It suffices therefore to prove that every element of the type $x=1+\sum_{n\ge 1} a_n X^n$ is invertible. But, for such an element x,

$|1-x|_p \le p^{-1}$, and therefore the series $\sum_{k=0}^{\infty}(1-x)^k$ converges in $\mathbb{S}_p$,

and $$|x\sum_{k=0}^{n}(1-x)^k-1|_p = |(1-x)^{n+1}|_p < p^{-n}.$$

This shows that $\sum_{k=0}^{\infty}(1-x)^k \in \mathbb{S}_p$ is the inverse of x. We have now proved that $\mathbb{S}_p$ is a *totally disconnected locally compact field of characteristic* p.

3. Locally compact totally disconnected fields. The fields $\mathbb{Q}_p$ and $\mathbb{S}_p$ are not the only examples of locally compact totally disconnected fields. However, all nondiscrete totally disconnected fields are finite algebraic extensions of $\mathbb{Q}_p$ or $\mathbb{S}_p$. A precise statement of this result will be given later without proofs . We shall now describe a locally compact field in general. For complete statements and proofs we refer to [W].

Observe first that every field is locally compact in the discrete topology. We exclude this trivial case and from now on we shall assume that $\mathfrak{F}$ is a *nondiscrete locally compact field.*

We let $\mathfrak{F}^*$ denote the multiplicative group of nonzero elements; let m be a translation-invariant measure on $\mathfrak{F}$, that is a nontrivial Borel measure m such that $m(E+x)=m(E)$ for every $x\in\mathfrak{F}$. If $x\in\mathfrak{F}^*$ then $m_x(E)=m(xE)$ is also translation-invariant and therefore, by the uniqueness of the Haar measure on locally compact groups [W], $m_x(E)=|x|m(E)$, where $|.|$ is a positive real number. The definition of $|.|$ shows that

(1) $|xy| = |x||y|$,

(2) $|.|$ *is a continuous function on* $\mathfrak{F}^*$.

We extend the definition of $|.|$ to all of $\mathfrak{F}$ by letting $|0|=0$. It is not difficult to show that

(3) $\{x\in\mathfrak{F}:\ |x|\le 1\}$ *is compact*,

and that indeed the sets $\{x\in\mathfrak{F}:\ |x|\le r\}$ are a base for the neighborhood system of zero. As a consequence of the compactness of $\{x\in\mathfrak{F}:\ |x|\le 1\}$ and the continuity of $|.|$, one obtains

$$|x+y| \le A \max(|x|,|y|),$$

with $A=\sup\{|x+1|: |x|\le 1\}$. The condition $A>1$ leads to the conclusion that $\mathfrak{F}=\mathbb{R}$ or $\mathfrak{F}=\mathbb{C}$. In the first case $|x|$ is the absolute value of $\mathbb{R}$, in the second case $|x|= a^2+ b^2$, where $x=a+ib$, $x\in\mathbb{C}$ and $a, b\in\mathbb{R}$.

We shall assume from now on that $A=1$, which implies the *ultrametric inequality* :

$$(4) \qquad |x+y| \le \max(|x|,|y|) \le |x| + |y|.$$

Observe that $|-1|^2=|(-1)^2|=1$, and therefore $|-x|=|x|$. This implies that $|x|=|(x+y)-y|\le\max(|x+y|,|y|)$. As a consequence

$$(5) \qquad |x+y| = |x| \quad \textit{if} \quad |y| < |x|.$$

Another consequence of the ultrametric inequality is that the set $B=\{|x|: x\in\mathfrak{F}^*\}$ is a discrete subgroup of $\mathbb{R}^+$. Indeed suppose that 1 is not an isolated point in B; then there exist $x_n\in\mathfrak{F}^*$ such that $|x_n|\ne 1$ for every n and $\lim |x_n|=1$. We may assume (taking if needed x_n^{-1}) that $|x_n|>1$ and that $|x_n|\le 2$.

Since $\{x: |x|\le 2\}$ is compact, there exists $a\in\mathfrak{F}$ such that $\lim x_n=a$. Clearly $\lim |x_n|=|a|$, that is $|a|=1$. But $\{a+x: |x|\le 1\}$ is a neighborhood of a. Therefore for n sufficiently large $x_n\in\{a+x: |x|\le 1\}$, and therefore by the ultrametric inequality $|x_n|\le 1$, contrary to the assumption that $|x_n|>1$. It follows that the sets $U_r=\{x: |x|\le r\}$ are open and compact. Therefore $\mathfrak{F}$ is *totally disconnected.* A discrete subgroup of $\mathbb{R}^+$ is generated by a positive number q. In other words, for each $x\in\mathfrak{F}^*$, there is $n\in\mathbb{Z}$ such that $|x|=q^n$ and

$$\{x: |x|\le q^n\}= \{x: |x|<q^{n+1}\}.$$

(3.1) *DEFINITION. Let* $\mathfrak{O}=\{x\in\mathfrak{F}: |x|\le 1\}$, $\mathfrak{O}_1=\{x\in\mathfrak{F}: |x|=1\}$ *and* $\mathfrak{p}=\{x\in\mathfrak{F}: |x|<1\}$.

We observe that $\mathfrak{O}$ is compact and open, that it is closed under addition and multiplication and that it contains all elements such that $|x^n|=|x|^n$ is bounded. It follows that $\mathfrak{O}$ is a *compact maximal subring* of $\mathfrak{F}$. If $x\in\mathfrak{O}$ is invertible in $\mathfrak{O}$ then $|x|\le 1$ and $|x^{-1}|=|x|^{-1}\le 1$, and therefore $x\in\mathfrak{O}_1$. Conversely, if $|x|=1$ then $|x^{-1}|=1$ and therefore x is invertible in $\mathfrak{O}$. Let $\{q^n: n\in\mathbb{Z}\}$ be the multiplicative subgroup of $\mathbb{R}^+$ onto which $|.|$ maps $\mathfrak{F}^*$, and suppose $p>1$. Let $\mathfrak{p}\in\mathfrak{O}$ be such that $|\mathfrak{p}|=q^{-1}$. Then $\mathfrak{P}=\mathfrak{p}\mathfrak{O}$, and $\mathfrak{P}$ is a unique maximal ideal of the ring $\mathfrak{O}$. It follows that $\mathfrak{O}/\mathfrak{P}=\mathfrak{f}$ is finite and, if we let m denote the Haar measure, $m(\mathfrak{P})=m(\mathfrak{p}\mathfrak{O})= |\mathfrak{p}|m(\mathfrak{O})=q^{-1}m(\mathfrak{O})$. Therefore q is a *positive integer which is exactly the index of* $\mathfrak{P}$ *in* $\mathfrak{O}$. Since $\mathfrak{O}/\mathfrak{P}$ is a finite field, $q=p^h$ for some prime p and positive integer h. An element $\mathfrak{p}\in\mathfrak{O}$ such that $|\mathfrak{p}|=q^{-1}$ is called a *prime element*.

Let $\xi_1,\xi_2,\dots,\xi_q$ be fixed elements of $\mathfrak{O}$ belonging to different classes of $\mathfrak{O}/\mathfrak{P}$. In other words $\xi_i\neq\xi_j$ mod $\mathfrak{P}$. We will show that every $x\in\mathfrak{F}$ can be expressed uniquely as a power series $x=\sum_{n\ge s} a_n\mathfrak{p}^n$ where $a_n\in\{\xi_1,\dots,\xi_q\}$. Indeed let $|x|=q^{-s}$; then $|\mathfrak{p}^{-s}x|=1$ and $\mathfrak{p}^{-s}=a_s$ mod $\mathfrak{P}$ where $a_s\in\{\xi_1,\xi_2,\dots,\xi_q\}$. In other words $|x-\mathfrak{p}^s a_s|\le q^{-s-1}$. Suppose that $a_s,a_{s+1},\dots,a_{s+k}$ have been found such that $|x-\sum_{n=s}^{s+k} a_{s+n}\mathfrak{p}^n|<q^{-s-k}$.

Then $\mathfrak{p}^{-s-k-1}(x-\sum_{n=s}^{s+k} a_{s+n}\mathfrak{p}^n)\in\mathfrak{O}$ and therefore there exists $a_{s+k+1}\in\{\xi_1,\xi_2,\dots,\xi_q\}$ such that $\mathfrak{p}^{-s-k-1}(x-\sum_{n=s}^{s+k} a_{s+n}\mathfrak{p}^n)=a_{s+k+1}$ mod $\mathfrak{P}$ which implies that $|x-\sum_{n=s}^{s+k+1} a_{s+n}\mathfrak{p}^n|<q^{-s-k-1}$. This proves that $x=\sum_{n\ge s} a_n\mathfrak{p}^n$.

We now state the structure theorem for locally compact totally disconnected field [W].

(3.2) *THEOREM. Let* $\mathfrak{F}$ *be a nondiscrete totally disconnected field. If* $\mathfrak{F}$ *has characteristic* $p>0$, *then* $\mathfrak{F}$ *is isomorphic to the field of power series* $\sum_{n\ge s} a_n X^n$ *with coefficients in a finite field of order* p^h. *The modulus of the series* $\sum_{n\geq s} a_n X^n$ *with* $a_s \neq 0$ *is* p^{hs}. *If* $\mathfrak{F}$ *has characteristic zero then* $\mathfrak{F}$ *is a finite algebraic extension of the p-adic field* $\mathbb{Q}_p$. *Let* q *be the order of the finite field* $\mathfrak{O}/\mathfrak{P}$ *and let* $\mathfrak{p}\in\mathfrak{P}$ *be such that* $|\mathfrak{p}|=q^{-1}$. *Then* $q=p^h$, *and* $\mathfrak{O}/\mathfrak{P}$ *is a finite extension of the field of order* p. *Every element of* $\mathfrak{F}$ *can be expressed as* $x=\sum_{n\geq s} a_n \mathfrak{p}^n$ *where the coefficients* a_n *are in* $\mathfrak{O}$ *and are uniquely determined modulo* $\mathfrak{P}$. *Finally* $|x|=q^{-s}$, *if* $a_s\neq 0$ *mod* $\mathfrak{P}$.

4. Two-dimensional lattices. Let $\mathfrak{F}$ be a locally compact totally disconnected field with modulus $|.|$.

Let $\mathfrak{O}=\{x\in\mathfrak{F}: |x|\leq 1\}$ and $\mathfrak{P}=\{x\in\mathfrak{F}: |x|<1\}$. Let $\mathfrak{p}\in\mathfrak{P}$ be such that $\mathfrak{p}\mathfrak{O}=\mathfrak{P}$. We let V denote a vector space of dimension 2 over $\mathfrak{F}$. Observe that V is locally compact and totally disconnected in the product topology. A subset of V is called a *module* over $\mathfrak{O}$ if it is a subgroup of V (with respect to addition) and is closed under multiplication by elements of $\mathfrak{O}$.

(4.1) *DEFINITION. A compact open subset of* V *which is a module over* $\mathfrak{O}$ *is called a lattice of* V.

If $\{e_1,e_2\}$ is a basis of V, then the set $e_1\mathfrak{O}\oplus e_2\mathfrak{O}$ is a lattice. A consequence of our next result is that every lattice is of this form. Observe that every lattice generates V as a vector space. Therefore it contains 2 linearly independent elements.

(4.2) *LEMMA. Let* $\mathcal{L}$ *be a lattice. Then the quotient* $\mathcal{L}/\mathfrak{p}\mathcal{L}$ *is a two-dimensional vector space over the finite field* $\mathfrak{f}=\mathfrak{O}/\mathfrak{P}$. *If* $e_1\in\mathcal{L}$ *and* $e_1\notin\mathfrak{p}\mathcal{L}$, *then there exists* $e_2\in\mathcal{L}$ *such that* $\mathcal{L}=e_1\mathfrak{O}\oplus e_2\mathfrak{O}$.

PROOF. The quotient $\mathcal{L}/\mathfrak{p}\mathcal{L}$ is a vector space over the finite field $\mathfrak{f}=\mathfrak{O}/\mathfrak{P}$. The dimension of this vector space cannot be 1. Indeed, if this were the case then $\mathcal{L}$ could be written as

$\mathcal{L}=e\mathfrak{O}+\mathfrak{p}\mathcal{L}$, for some nonzero $e\in\mathcal{L}$. Inductively one could then write $\mathcal{L}=e\mathfrak{O}+\mathfrak{p}^k\mathcal{L}$, which implies, since k is arbitrarily large, that $\mathcal{L}=e\mathfrak{O}$, contradicting the assumption that $\mathcal{L}$ is open in V. This implies that $\mathcal{L}/\mathfrak{p}\mathcal{L}$ is of dimension 2. As a consequence there exists $e_2\in\mathcal{L}$ such that $\{e_1+\mathfrak{p}\mathcal{L}, e_2+\mathfrak{p}\mathcal{L}\}$ is a basis for $\mathcal{L}/\mathfrak{p}\mathcal{L}$. Therefore $\mathcal{L}=e_1\mathfrak{O}+e_2\mathfrak{O}+\mathfrak{p}\mathcal{L}$ and, again by induction, $\mathcal{L}=e_1\mathfrak{O}+e_2\mathfrak{O}+\mathfrak{p}^k\mathcal{L}$, which implies $\mathcal{L}=e_1\mathfrak{O}\oplus e_2\mathfrak{O}$. ∎

The next result is the two-dimensional version of a theorem valid for n-dimensional vector spaces over $\mathfrak{F}$, which is known as the *theorem of elementary divisors*. The proof in the general case may be obtained using induction on the dimension of the vector space.

(4.3) *THEOREM. Let $\mathcal{L}$ and $\mathcal{L}'$ be two lattices of V. Then there exist vectors e_1 and e_2 and integers* n *and* m *such that $\mathcal{L}= e_1\mathfrak{O}\oplus e_2\mathfrak{O}$, and $\mathcal{L}'=e_1\mathfrak{p}^n\mathfrak{O}\oplus e_2\mathfrak{p}^m\mathfrak{O}$. Moreover the set* $\{n,m\}$ *depends only on $\mathcal{L}$ and $\mathcal{L}'$.*

PROOF. Since $\mathcal{L}$ is an open set containing the origin and $\mathfrak{p}^k\mathcal{L}'$ is a basis of neighborhoods of the origin, for some integer j, $\mathfrak{p}^j\mathcal{L}'\subseteq\mathcal{L}$. Let j be the smallest such integer, and define $\mathcal{L}''=\mathfrak{p}^j\mathcal{L}'$. Then $\mathcal{L}''\subseteq\mathcal{L}$, but $\mathcal{L}''$ is not contained in $\mathfrak{p}\mathcal{L}$. Let e_1 be an element of $\mathcal{L}''$ which is not in $\mathfrak{p}\mathcal{L}$. Then, by (4.2), there exists e_2, such that $\mathcal{L}=e_1\mathfrak{O}\oplus e_2\mathfrak{O}$. Let i be the smallest integer such that $\mathfrak{p}^i e_2\in\mathcal{L}''$. We assert that $\mathcal{L}''=e_1\mathfrak{O}\oplus e_2\mathfrak{p}^i\mathfrak{O}$. Indeed, if $v\in\mathcal{L}''\subseteq \mathcal{L}=e_1\mathfrak{O}\oplus e_2\mathfrak{O}$, then $v=\alpha e_1+\beta e_2$, with $\alpha,\beta\in\mathfrak{O}$. But $\alpha e_1\in\mathcal{L}''$, and therefore $\beta e_2\in\mathcal{L}''$. Let $\beta=\lambda\mathfrak{p}^s$, with $|\lambda|=1$. Then $\mathfrak{p}^s e_2\in\mathcal{L}''$, which implies $s\geq i$. Therefore $\beta e_2=\lambda\mathfrak{p}^{s-i}\mathfrak{p}^i e_2\in\mathfrak{p}^i\mathfrak{O}$. We conclude that $\mathcal{L}''=e_1\mathfrak{O}\oplus e_2\mathfrak{p}^i\mathfrak{O}$, and $\mathcal{L}=e_1\mathfrak{p}^n\mathfrak{O}\oplus e_2\mathfrak{p}^m\mathfrak{O}$, with $n=-j$ and $m=i-j$. Observe now that $-\min(n,m)$ is the smallest integer j such that $\mathfrak{p}^j\mathcal{L}'\subseteq\mathcal{L}$, and that $q^{|n-m|}$ is the index of $\mathfrak{p}^j\mathcal{L}'$ in $\mathcal{L}$. Therefore the set $\{n,m\}$ is uniquely determined by $\mathcal{L}$ and $\mathcal{L}'$. ∎

5. The tree of PGL(2,$\mathfrak{F}$). Let V be a fixed two-dimensional vector space over $\mathfrak{F}$. If $\mathcal{L}$ and $\mathcal{L}'$ are lattices in V, we say that $\mathcal{L}$ and $\mathcal{L}'$ are *equivalent* when there exists $\alpha\in\mathfrak{F}^*$ such that $\alpha\mathcal{L}=\mathcal{L}'$. The set of equivalence classes of lattices will be denoted by $\mathfrak{X}$. Given two elements Λ and Λ' of $\mathfrak{X}$, and two lattices $\mathcal{L}\in\Lambda$ and $\mathcal{L}'\in\Lambda'$, by (4.3) there exists a basis $\{e_1,e_2\}\in V$ and two integers n, m such that $\mathcal{L}=e_1\mathfrak{O}\oplus e_2\mathfrak{O}$, and $\mathcal{L}'=\mathfrak{p}^n e_1\mathfrak{O}\oplus\mathfrak{p}^m e_2\mathfrak{O}$. Clearly $\mathcal{L}'\subseteq\mathcal{L}$ if and only if $n,m\geq 0$. We observe that the integer $|n-m|$ depends only on the equivalence classes Λ and Λ'. Indeed, if $x,y\in\mathfrak{F}^*$, then $\{xe_1,xe_2\}$ is a basis for $x\mathcal{L}$ and $\{y\mathfrak{p}^n e_1,y\mathfrak{p}^m e_2\}$ is a basis for $y\mathcal{L}'$. But $y\mathfrak{p}^n e_1\mathfrak{O}\oplus y\mathfrak{p}^m e_2\mathfrak{O}=\mathfrak{p}^{s+n}xe_1\mathfrak{O}\oplus\mathfrak{p}^{s+m}xe_2\mathfrak{O}$, where $q^{-s}=|yx^{-1}|$. We shall say that the number $|n-m|=\mathbf{d}(\Lambda,\Lambda')$ is the *distance* between Λ and Λ'. We have not shown yet that $\mathbf{d}$ even defines a metric. We shall prove instead that the graph structure introduced in $\mathfrak{X}$ by the following definition yields a tree. The fact that $\mathbf{d}$ is a metric will then follow immediately.

(5.1) *DEFINITON. Let $\Lambda,\Lambda'\in\mathfrak{X}$; we say that $\{\Lambda,\Lambda'\}$ is an edge or that $\{\Lambda,\Lambda'\}\in\mathfrak{E}$, if $\mathbf{d}(\Lambda,\Lambda')=1$.*

Our next task is to show that $(\mathfrak{X},\mathfrak{E})$ is a tree. We need a result which is essentially nothing but a convenient reformulation of (4.3).

(5.2) *LEMMA. Let $\mathcal{L}_0$ be a lattice in V, and Λ an element of $\mathfrak{X}$. Suppose that Λ' is another element of $\mathfrak{X}$, and $\Lambda'\neq\Lambda$; then the following conditions are equivalent for a lattice $\mathcal{L}\in\Lambda$.*

(1) $\mathcal{L}\subseteq\mathcal{L}_0$, and for some $n>0$ and some basis $\{e_1,e_2\}$ $\mathcal{L}=\mathfrak{p}^n e_1\mathfrak{O}\oplus e_2\mathfrak{O}\subseteq e_1\mathfrak{O}\oplus e_2\mathfrak{O}=\mathcal{L}_0$.

(2) $\mathcal{L}\subseteq\mathcal{L}_0$ and $\mathcal{L}\not\subseteq\mathfrak{p}\mathcal{L}_0$.

(3) $\mathcal{L}$ is the maximal sublattice of $\mathcal{L}_0$ which belongs to Λ'.

(4) $\mathcal{L}_0/\mathcal{L}$ is generated by one element.

Moreover there exists only one $\mathcal{L}\in\Lambda$ satisfying these conditions.

PROOF. It is clear that (1) implies (2). If (2) holds and $\mathcal{L}'=\alpha\mathcal{L}$, with $\mathcal{L}\subseteq\alpha\mathcal{L}\subseteq\mathcal{L}_0$, then $\alpha=\beta\mathfrak{p}^k$, with $|\beta|=1$, and $\mathcal{L}'=\mathfrak{p}^k\mathcal{L}$. Since $\mathcal{L}\not\subset\mathfrak{p}\mathcal{L}_0$ and $\mathcal{L}\not\subset\mathfrak{p}\mathcal{L}$, we must have k=0. Thus (2) implies (3). Suppose (3) holds and let $\{e_1,e_2\}$ be a basis chosen according to (4.3) so that $\mathcal{L}_0=e_1\mathfrak{O}\oplus e_2\mathfrak{O}$ and $\mathcal{L}=\mathfrak{p}^ke_1\mathfrak{O}\oplus\mathfrak{p}^he_2\mathfrak{O}$. Since $\mathcal{L}\subseteq\mathcal{L}_0$, we must have h, k≥0. If j=min(h,k), then $\mathcal{L}\subseteq\mathfrak{p}^j\mathcal{L}_0\subseteq\mathcal{L}_0$. This implies j=0 and (1) follows. Thus (3) implies (1). Now (4) is clearly implied by (1). On the other hand if (4) holds $\mathcal{L}$ cannot be contained in $\mathfrak{p}\mathcal{L}_0$, because $\mathcal{L}_0/\mathfrak{p}\mathcal{L}_0$ is two-dimensional. The existence of a lattice of Λ satisfying (1) follows directly from (4.3). It suffices to observe that any lattice of Λ has the form $\mathfrak{p}^ne_1\mathfrak{O}\oplus\mathfrak{p}^me_2\mathfrak{O}$, where $\mathcal{L}_0=e_1\mathfrak{O}\oplus e_2\mathfrak{O}$, and therefore, if n≥m, $\mathfrak{p}^{n-m}e_1\mathfrak{O}\oplus\mathfrak{p}e_2\mathfrak{O}$ satisfies (1). Uniqueness is also easy for a lattice of Λ satisfying (2). Indeed if $\mathcal{L}$ and $\mathcal{L}'=\alpha\mathcal{L}$ both satisfy (2), it follows that $|\alpha|=1$.∎

We observe that, if $\mathcal{L}'$ satisfies one of the conditions of the lemma, then $\mathcal{L}/\mathcal{L}'$ is isomorphic to $\mathfrak{O}/\mathfrak{p}^n\mathfrak{O}$, where $n=\mathbf{d}(\Lambda,\Lambda')$. In particular $\mathbf{d}(\Lambda,\Lambda')=0$ if and only if $\Lambda=\Lambda'$; furthermore $\mathbf{d}(\Lambda,\Lambda')=1$ if and only if there exist $\mathcal{L}\in\Lambda$ and $\mathcal{L}'\in\Lambda'$, such that $\mathcal{L}'\subseteq\mathcal{L}$ and $\mathcal{L}/\mathcal{L}'$ is isomorphic to the field $\mathfrak{O}/\mathfrak{p}\mathfrak{O}$ of order q.

(5.3) *THEOREM . The graph $\mathfrak{X}$ is a tree .*

PROOF. We show first that $\mathfrak{X}$ is connected. Let $\Lambda,\Lambda'\in\mathfrak{X}$, and let $\mathcal{L}\in\Lambda$ and $\mathcal{L}'\in\Lambda'$ be such that $\mathcal{L}'\subseteq\mathcal{L}$ and $\mathcal{L}'\not\subset\mathfrak{p}\mathcal{L}$. By (5.2), there exists a basis $\{e_1,e_2\}$ such that $\mathcal{L}=e_1\mathfrak{O}\oplus e_2\mathfrak{O}$ and $\mathcal{L}'=\mathfrak{p}^he_1\mathfrak{O}\oplus e_2\mathfrak{O}$. For i=0,1,...,h, let $\mathcal{L}_i=\mathfrak{p}^ie_1\mathfrak{O}\oplus e_2\mathfrak{O}$. If Λ_i is the class of $\mathcal{L}_i$, we have that $\mathbf{d}(\Lambda_i,\Lambda_{i+1})=1$, for i=0,1,...,h. Therefore $\{\Lambda=\Lambda_0,\Lambda_1,\ldots$ $\ldots,\Lambda_h=\Lambda'\}$ is a path in the graph connecting Λ and Λ'. We show now that $\mathfrak{X}$ contains no circuits. We will prove the following statement (which directly implies the absence of circuits): if $\Lambda_0,\Lambda_1,\ldots,\Lambda_h$ is a sequence such that $\mathbf{d}(\Lambda_i,\Lambda_{i+1})=1$ for every i, and $\Lambda_i\neq\Lambda_{i-2}$ for i=2,...,h-2, then $\mathbf{d}(\Lambda_0,\Lambda_h)=h$. This statement is obvious if h=0 or 1. We make the induction hypothesis that it is true for h-1. Choose $\mathcal{L}_0 \in \Lambda_0$, and let $\mathcal{L}_i$ be the maximal element of Λ_i contained in $\mathcal{L}_{i-1}$. In other words, $\mathcal{L}_0 \supseteq \mathcal{L}_1 \supseteq \ldots \supseteq \mathcal{L}_n$ and $\mathcal{L}_i \not\subset \mathfrak{p}\mathcal{L}_{i-1}$. This means that $\mathcal{L}_i$ has index q in $\mathcal{L}_{i-1}$ and therefore $\mathcal{L}_j$ has index q^j in $\mathcal{L}_0$, for every j. Now $\mathbf{d}(\Lambda_0, \Lambda_{h-1}) = h-1$, by induction hypothesis, and $\mathcal{L}_{h-1} \subseteq \mathcal{L}_0$ with index q^{h-1}. This means that if, according to (4.3), $\mathcal{L}_{h-1} = \mathfrak{p}^n e_1 \mathfrak{O} \oplus \mathfrak{p}^m e_2 \mathfrak{O}$, with $\mathcal{L}_0 = e_1 \mathfrak{O} \oplus e_2 \mathfrak{O}$, it must be the case that $n, m \geq 0$, and $n+m = h-1 = |n-m|$. It follows that either n or m is zero, and $\mathcal{L}_{h-1}$ satisfies the conditions of (5.2), in particular $\mathcal{L}_{h-1} \not\subset \mathfrak{p}\mathcal{L}_0$. Since $\mathcal{L}_{h-1}/\mathcal{L}_h$ is a field of order q, $\mathcal{L}_{h-1}$ is generated by $\mathcal{L}_h$ and a one- dimensional subspace. But also $\mathfrak{p}\mathcal{L}_{h-2} \subseteq \mathcal{L}_{h-1}$ and (5.2) implies that $\mathcal{L}_{h-1}$ is generated by $\mathfrak{p}\mathcal{L}_{h-2}$ and a one-dimensional subspace. In other words $\mathcal{L}_h$ and $\mathfrak{p}\mathcal{L}_{h-2}$ are inverse images of two one-dimensional subspaces of the finite plane $\mathcal{L}_{h-1}/\mathfrak{p}\mathcal{L}_{h-1}$, with respect to the field $\mathfrak{f} = \mathfrak{O}/\mathfrak{p}\mathfrak{O}$. But $\mathfrak{p}\mathcal{L}_{h-2} \neq \mathcal{L}_h$, because otherwise $\Lambda_h = \Lambda_{h-2}$, contrary to our assumption. Therefore $\mathcal{L}_{h-1} = \mathcal{L}_h + \mathfrak{p}\mathcal{L}_{h-2}$. It follows that $\mathcal{L}_h \not\subset \mathfrak{p}\mathcal{L}_0$, because otherwise $\mathcal{L}_{h-1} \subseteq \mathfrak{p}\mathcal{L}_0$. We have proved therefore that $\mathcal{L}_h$ satisfies the conditions of (5.2) with reference to $\mathcal{L}_0$. This means that the distance of Λ_h and Λ_0 is the logarithm in base q of the index of $\mathcal{L}_h$ in $\mathcal{L}$. Therefore $\mathbf{d}(\Lambda_0, \Lambda_h) = h$. ∎

The proof of (5.3) shows that $\mathbf{d}(\Lambda, \Lambda')$ coincides with the distance of Λ and Λ' as vertices of the tree $\mathfrak{X}$. This implies, of course, that $\mathbf{d}$ satisfies the triangle inequality. We shall prove now that $\{\Lambda: \mathbf{d}(\Lambda, \Lambda_0) = 1\}$ consists of q+1 elements. We known, by (5.2), that if $\mathcal{L}_0 \in \Lambda_0$ there exists one and only one $\mathcal{L} \in \Lambda$ such that $\mathcal{L} \subseteq \mathcal{L}_0$ and $\mathcal{L} \not\subset \mathfrak{p}\mathcal{L}_0$. Therefore the classes Λ with $\mathbf{d}(\Lambda, \Lambda_0) = 1$ are just as many as the lattices contained in $\mathcal{L}_0$ and satisfying the conditions of (5.2), in fact, by ((5.2),2)), as many as the lattices $\mathcal{L}$ such that $\mathfrak{p}\mathcal{L}_0 \subseteq \mathcal{L} \subseteq \mathcal{L}_0$ and $\mathfrak{p}\mathcal{L}_0 \neq \mathcal{L} \neq \mathcal{L}_0$. Every such lattice is in one-to-one correspondence with a one-dimensional subspace of $\mathcal{L}_0/\mathfrak{p}\mathcal{L}_0$, which is a two-dimensional vector space over the finite field $\mathfrak{f}=\mathfrak{O}/\mathfrak{p}\mathfrak{O}$ of q elements.

The lines passing through the origin in the finite plane $(\mathfrak{O}/\mathfrak{p}\mathfrak{O})\times(\mathfrak{O}/\mathfrak{p}\mathfrak{O})$ are as many as the points in the projective line over the field $\mathfrak{O}/\mathfrak{p}\mathfrak{O}$, that is q+1. This shows that there are q+1 distinct Λ such that $\mathbf{d}(\Lambda,\Lambda_0)=1$. Let now $\mathcal{L}_0=\mathfrak{O}\times\mathfrak{O}$ be the lattice containing the canonical vectors (1,0) and (0,1), and Λ_0 be the class to which $\mathcal{L}_0$ belongs. We shall prove that the infinite chains $\{\Lambda_0,\Lambda_1,\ldots,\Lambda_n,\ldots\}$ are in one-to-one correspondence with the lines in the plane $\mathfrak{F}\times\mathfrak{F}=V$.

(5.4) *PROPOSITION. Let* $\mathcal{R}$ *be the set of one-dimensional subspaces of* V, *and let* $\omega\in\Omega$ *be a point of the boundary of* $\mathfrak{X}$ *associated to the infinite chain* $\{\Lambda_0,\Lambda_1,\ldots,\Lambda_n,\ldots\}$. *Let* $\mathcal{L}_i\in\Lambda_i$ *be such that* $\mathcal{L}_j\subseteq\mathcal{L}_{j-1}$ *and* $\mathcal{L}_j\not\subset\mathfrak{p}\mathcal{L}_{j-1}$. *Then* $\bigcap_{j=0}^{\infty}\mathcal{L}_j$ *generates a one-dimensional subspace* r *of* V *and the correspondence* $\omega\longrightarrow r$ *is a bijection between* Ω *and* $\mathcal{R}$.

PROOF. We will prove that the inverse correspondence $r\longrightarrow\vartheta(r)=\omega$ is a bijection between $\mathcal{R}$ and Ω. Let $r\in\mathcal{R}$; we consider the following sequence of lattices $L_j=r\cap\mathcal{L}_0+\mathfrak{p}^j\mathcal{L}_0$. There exists a vector $v_0\in r\cap\mathcal{L}_0$ such that $r\cap\mathcal{L}_0=\mathfrak{O}v_0$. Indeed, it is obvious that $\mathfrak{O}v\subseteq r\cap\mathcal{L}_0$ for every $v\in r\cap\mathcal{L}_0$; observe further that, if $v,w\in r\cap\mathcal{L}_0$ and $v=\alpha w$ with $\alpha\in\mathfrak{F}$, then $\mathfrak{O}v\subseteq\mathfrak{O}w$ if and only if $|\alpha|\leq1$. Therefore all the sets $\mathfrak{O}v$ with $v\in r\cap\mathcal{L}_0$ form a chain under inclusion. This means by the compactness of $r\cap\mathcal{L}_0$ that there exists a maximal set $\mathfrak{O}v_0$ such that $r\cap\mathcal{L}_0=\mathfrak{O}v_0$. Observe that $v_0\notin\mathfrak{p}\mathcal{L}_0$. By (4.2) there exists $w_0\in\mathcal{L}_0\setminus\mathfrak{p}\mathcal{L}_0$ such that $\mathcal{L}_0=\mathfrak{O}v_0\oplus\mathfrak{O}w_0$. This means that $L_j=\mathfrak{O}v_0\oplus\mathfrak{p}^j\mathfrak{O}w_0$; therefore it is clear that $L_{j+1}\subseteq L_j$, $L_0=\mathcal{L}_0$ and that L_{j+1} has index q in L_j. In particular L_{j+1} is maximal in L_j and if $\Lambda_j=[L_j]$ then $\mathbf{d}(\Lambda_j,\Lambda_{j+1})=1$ for every j. To show that $\{\Lambda_0,\Lambda_1,\ldots,\Lambda_j,\ldots\}$ is an infinite chain it suffices to prove that $\Lambda_{j+2}\neq\Lambda_j$ for every j. But if $\Lambda_{j+2}=\Lambda_j$ then, for some integer k, $\mathfrak{p}^kL_{j+2}=L_j$, that is $\mathfrak{p}^k\mathfrak{O}v_0\oplus\mathfrak{p}^{k+j+2}\mathfrak{O}w_0=\mathfrak{O}v_0\oplus\mathfrak{p}^j\mathfrak{O}w_0$. This is a contradiction because $\mathfrak{p}^kv_0\in L_j$ implies that $\mathfrak{p}^k\in\mathfrak{O}$, that is $k\geq0$, while $v_0 \in \mathfrak{p}^k L_{j+2}$ implies that $\mathfrak{p}^{-k} \in \mathfrak{O}$, that is $k \le 0$. Therefore $k=0$ and $L_j = L_{j+2}$, a contradiction. This proves that $\vartheta(r) = \{\Lambda_0, \Lambda_1, \dots, \Lambda_j, \dots\}$ is a chain. We observe that $\bigcap_{j=0}^{\infty} L_j = \mathfrak{O}v_0 = r \cap \mathcal{L}_0$, and therefore the linear subspace generated by $\bigcap_{j=0}^{\infty} L_j$ is r. We prove now that the map $r \longmapsto \vartheta(r)$ is injective. Indeed if $\vartheta(r) = \vartheta(r')$ then $L_j = r \cap \mathcal{L}_0 + \mathfrak{p}^j \mathcal{L}_0$ is equivalent to $L'_j = r' \cap \mathcal{L}_0 + \mathfrak{p}^j \mathcal{L}_0$ for every j. Since L_j and L'_j are sublattices of $\mathcal{L}_0$ both of index q^j and equivalent, it follows that $L'_j = L_j$ for every j. In particular $\bigcap_{j=0}^{\infty} L_j = \bigcap_{j=0}^{\infty} L'_j$ and so $r = r'$. Finally we prove that ϑ is surjective. Let ω be in Ω and $\omega = \{\Lambda_0, \Lambda_1, \dots, \Lambda_j, \dots\}$. Let $\mathcal{L}_j$ be the maximal element of Λ_j contained in $\mathcal{L}_{j-1}$ for every $j=1,2,\dots$. We prove now that $\bigcap_{j=0}^{\infty} \mathcal{L}_j$ generates a line $r \in \mathcal{R}$. Indeed $\mathcal{L}_j$ has index q in $\mathcal{L}_{j-1}$ because $\mathbf{d}(\Lambda_j, \Lambda_{j-1}) = 1$ and so $\mathcal{L}_j$ has index q^j in $\mathcal{L}_0$. Since $\mathbf{d}(\Lambda_0, \Lambda_j) = j$ this implies that $\mathcal{L}_j$ is the maximal element of Λ_j contained in $\mathcal{L}_0$. In particular $\mathcal{L}_j \not\subset \mathfrak{p}\mathcal{L}_0$. Let $v_j \in \mathcal{L}_j \setminus \mathfrak{p}\mathcal{L}_0$. Passing to a sub-sequence we may assume that v_j converges. Let $\lim_j v_j = v_0$; then $v_0 \in \mathcal{L}_0 \setminus \mathfrak{p}\mathcal{L}_0$ and $v_0 \ne 0$. Since $v_k \in \mathcal{L}_j$ for $k>j$ and $\mathcal{L}_j$ is closed it follows that $v_0 \in \mathcal{L}_j$ for every j and so $v_0 \in \bigcap_{j=0}^{\infty} \mathcal{L}_j \ne 0$. But $\bigcap_{j=0}^{\infty} \mathcal{L}_j$ is a module with respect to $\mathfrak{O}$; since $\mathcal{L}_j$ has index q^j in $\mathcal{L}_0$ the index of $\bigcap_{j=0}^{\infty} \mathcal{L}_j$ in $\mathcal{L}_0$ must be infinite. Therefore by (4.1) $\bigcap_{j=0}^{\infty} \mathcal{L}_j$ cannot contain a basis, and the subspace r it generates must be one-dimensional. Also we observe that $v_0 \in \mathcal{L}_j$ for every j and $\mathfrak{O}v_0 = r \cap \mathcal{L}_0$ because $v_0 \in (\mathcal{L}_0 \cap r) \setminus \mathfrak{p}\mathcal{L}_0$. We prove now that $\vartheta(r) = \omega$; indeed $\mathcal{L}_0 = \mathfrak{O}v_0 \oplus \mathfrak{O}w_0$ for some $w_0 \in \mathcal{L}_0 \setminus \mathfrak{p}\mathcal{L}_0$, and $\vartheta(r) = \{\Lambda_0, [L_1], [L_2], \dots, [L_j], \dots\}$ where $L_j = \mathfrak{O}v_0 \oplus \mathfrak{p}^j \mathfrak{O}w_0$. Since $\mathcal{L}_j \subseteq \mathcal{L}_0 = \mathfrak{O}v_0 \oplus \mathfrak{O}w_0$ and $v_0 \in \mathcal{L}_j$, there exists a k such that

$\mathfrak{p}^k w_0 \in \mathcal{L}_j \setminus \mathfrak{p}\mathcal{L}_j$. It follows that $\mathcal{L}_j = \mathfrak{O} v_0 \oplus \mathfrak{p}^k \mathfrak{O} w_0$. Because the index of $\mathcal{L}_j$ in $\mathcal{L}_0$ is equal to q^j we have that $k=j$ and $\mathcal{L}_j = L_j$ for every j. This means that $\vartheta(\kappa)=\omega$ and the proposition follows. ▮

We consider the following groups of two-by-two matrices with entries in $\mathfrak{F}$: **GL**$(2,\mathfrak{F})$ is the group of nonsingular matrices; **PGL**$(2,\mathfrak{F})$ is the quotient of **GL**$(2,\mathfrak{F})$ modulo its center which consists of nonzero scalar multiples of the identity; **PSL**$(2,\mathfrak{F})$ is the group of matrices with determinant 1 modulo $\{+I,-I\}$. The groups **GL**$(2,\mathfrak{F})$, **PGL**$(2,\mathfrak{F})$ and **PSL**$(2,\mathfrak{F})$ are locally compact groups when endowed with the usual topology. Choosing the matrix entries in the ring $\mathfrak{O}$ we obtain the subgroups **GL**$(2,\mathfrak{O})$, **PGL**$(2,\mathfrak{O})$ and **PSL**$(2,\mathfrak{O})$ which are open and compact, in **GL**$(2,\mathfrak{F})$, **PGL**$(2,\mathfrak{F})$ and **PSL**$(2,\mathfrak{F})$, respectively.

We will describe now the action of **PGL**$(2,\mathfrak{F})$ and **PSL**$(2,\mathfrak{F})$ on the tree $\mathfrak{X}$. We fix a basis $\{e_1,e_2\} \subseteq \mathcal{V}$, and let $\mathcal{L}_0$ be the lattice generated by this basis, and Λ_0 its equivalence class. With respect to the basis $\{e_1,e_2\}$ each nonsingular matrix A defines a linear transformation on $\mathcal{V}$ which maps a lattice into a lattice. Furthermore, if $\alpha \in \mathfrak{F}^*$, then $A(\alpha\mathcal{L}) = \alpha A(\mathcal{L})$. Therefore if $\mathcal{L}$ and $\mathcal{L}'$ are equivalent so are $A(\mathcal{L})$ and $A(\mathcal{L}')$; in other words A acts on the equivalence classes of lattices and therefore on $\mathfrak{X}$.

Suppose now that Λ, $\Lambda' \in \mathfrak{X}$ and $\mathbf{d}(\Lambda,\Lambda')=n$. Assume that $\mathcal{L} \in \Lambda$ and $\mathcal{L}' \in \Lambda'$ is the maximal element of Λ' contained in $\mathcal{L}$. Then $A(\mathcal{L}') \subseteq A(\mathcal{L})$ and $A(\mathcal{L}')$ is maximal in its class with respect to this property. Therefore $\mathbf{d}(A(\mathcal{L}'),A(\mathcal{L}))=n$. Thus A is an isometry of $\mathfrak{X}$. It follows that every nonsingular matrix defines an automorphism of $\mathfrak{X}$. Observe however that, if αI, with $\alpha \in \mathfrak{F}^*$, is a nonzero multiple of the identity matrix, then $\alpha I(\mathcal{L}) = \alpha\mathcal{L}$ maps each element of $\mathfrak{X}$ into itself. Conversely, if A is a matrix which acts as the identity on $\mathfrak{X}$, then $A=\alpha I$ for some $\alpha \in \mathfrak{F}^*$. Therefore **PGL**$(2,\mathfrak{F})$=**GL**$(2,\mathfrak{F})/\{\alpha I:\ \alpha \in \mathfrak{F}^*\}$ is isomorphic to a subgroup of $\mathrm{Aut}(\mathfrak{X})$. It is not difficult to see that this subgroup is closed and that the natural topology of **PGL**$(2,\mathfrak{F})$ is the same as its relative topology in $Aut(\mathfrak{X})$.

We will now describe the action of $\mathbf{PGL}(2,\mathfrak{F})$ on the boundary of $\mathfrak{X}$. By (5.4) the boundary Ω of $\mathfrak{X}$ is in one-to-one correspondence with the set $\mathcal{R}$ of one-dimensional vector subspaces of V. The set $\mathcal{R}$ has a natural compact topology once it is identified with the *projective line* relative to $\mathfrak{F}$, i.e. the space of equivalence classes of nonzero vectors (α,β), two vectors being equivalent when one is a nonzero multiple of the other. The group $\mathbf{GL}(2,\mathfrak{F})$ acts naturally on the projective line, because it maps every nonzero vector into a nonzero vector and preserves equivalence classes. Therefore $\mathbf{PGL}(2,\mathfrak{F})$ is a group of continuous trasformations of $\mathcal{R}$, indeed the group of *projective transformations*. Since $\mathbf{PGL}(2,\mathfrak{F})$ acts continuosly on Ω it is natural to ask whether Ω and $\mathcal{R}$ have the same topology and whether the action of $\mathbf{PGL}(2,\mathfrak{F})$ is the same. The answer is affirmative on both counts. The correspondence

$$r \longrightarrow \{[r \cap \mathcal{L}_0 + p^n\mathcal{L}_0]:\ n \in \mathbb{N}\}$$

is continuous, and therefore bicontinuous, and $A(r)$ corresponds to the sequence of lattices $\{[A(r)\cap A(\mathcal{L}_0) + p^nA(\mathcal{L}_0)]\} = \{A([r\cap\mathcal{L}_0+p^n\mathcal{L}_0])\}$. Observe that $\mathbf{PGL}(2,\mathfrak{F})$ acts transitively on $\mathfrak{X}$, but also on $\mathcal{R}$ and therefore on Ω.

The group $\mathbf{PSL}(2,\mathfrak{O})$ of matrices with entries in $\mathfrak{O}$ and determinant 1 is exactly the subgroup of $\mathbf{GL}(2,\mathfrak{F})$ which maps $\mathcal{L}_0$ onto itself. Thus $\mathbf{PSL}(2,\mathfrak{O})$ fixes Λ_0 and is a closed subgroup of the group K_{Λ_0}. We show that $\mathbf{PSL}(2,\mathfrak{O})$ acts transitively on Ω. In view of the remarks above it suffices to show that $\mathbf{PSL}(2,\mathfrak{O})$ acts transitively on the projective line $\mathcal{R}$. This is obvious because every element of $\mathcal{R}$ is determined by a column vector with one entry in $\mathfrak{O}$ and the other entry equal to the identity of $\mathfrak{F}$. It is not difficult to show that $\mathbf{PGL}(2,\mathfrak{F})$ is a proper subgroup of $Aut(\mathfrak{X})$.

Indeed the identification of Ω with the projective line $\mathcal{R}$ and the resulting identification of the action of $\mathbf{PGL}(2,\mathfrak{F})$ on Ω

with its natural action on $\mathcal{R}$ imply that every element of $\mathbf{PGL}(2,\mathfrak{F})$ which fixes three points of $\mathcal{R}$ is the identity. On the other hand we can find proper rotations in $\mathrm{Aut}(\mathfrak{X})$ which fix every element of any finite set. This shows that the inclusions of $\mathbf{PGL}(2,\mathfrak{F})$ and $\mathbf{PSL}(2,\mathfrak{O})$ in $\mathrm{Aut}(\mathfrak{X})$ and K_{Λ_0}, respectively, are proper.

We turn now to consider the subgroup $\mathbf{PSL}(2,\mathfrak{F})$ which containing $\mathbf{PSL}(2,\mathfrak{O})$ acts transitively on Ω. We observe that $\mathbf{PSL}(2,\mathfrak{F})$ is noncompact and therefore by (I,10.2) it has at most two orbits on $\mathfrak{X}$. We shall prove that the orbits are exactly two by showing that no element of $\mathbf{PSL}(2,\mathfrak{F})$ can map $e_1\mathfrak{O}\oplus e_2\mathfrak{O}=\mathcal{L}_0$ onto a multiple of $\mathcal{L}_1=\mathfrak{p}e_1\mathfrak{O}\oplus e_2\mathfrak{O}\subseteq\mathcal{L}_0$. If $A=\begin{bmatrix} a & b \\ c & d \end{bmatrix}$ is a matrix then $A(\mathcal{L}_0)\subseteq\mathcal{L}_1$ if and only if $a,b\in\mathfrak{p}\mathfrak{O}$ and $c,d\in\mathfrak{O}$, while $A(\mathcal{L}_1)\subseteq\mathcal{L}_0$ if and only if $a,c\in\mathfrak{p}^{-1}\mathfrak{O}$ and $b,d\in\mathfrak{O}$. If $A\in\mathbf{PSL}(2,\mathfrak{F})$ and $A(\mathcal{L}_0)$ is a multiple of $\mathcal{L}_1$, we must have, for some $n\in\mathbb{Z}$, $\mathfrak{p}^nA(\mathcal{L}_0)=\mathcal{L}_1$, which implies $\mathfrak{p}^nA(\mathcal{L}_0)\subseteq\mathcal{L}_1$ and $\mathfrak{p}^{-n}A^{-1}(\mathcal{L}_1)\subseteq\mathcal{L}_0$. These conditions imply that, if $A=\begin{bmatrix} a & b \\ c & d \end{bmatrix}$, then $\mathfrak{p}^{n-1}a,\mathfrak{p}^{n-1}b\in\mathfrak{O}$; $\mathfrak{p}^nc,\mathfrak{p}^nd\in\mathfrak{O}$. This implies that $\mathfrak{p}^{2n-1}(ad-bc)=\mathfrak{p}^{2n-1}\in\mathfrak{O}$, which means that $n>0$. On the other hand $\mathfrak{p}^{-n}A^{-1}(\mathcal{L}_1)\subset\mathcal{L}_0$ implies that $\mathfrak{p}^{1-n}d$, $\mathfrak{p}^{1-n}c\in\mathfrak{O}$ and $\mathfrak{p}^{-n}b,\mathfrak{p}^{-n}a\in\mathfrak{O}$; this means that $\mathfrak{p}^{1-2n}(ad-bc)=\mathfrak{p}^{1-2n}\in\mathfrak{O}$ and so $n\leq 0$, a contradiction. It follows that $\mathbf{PSL}(2,\mathfrak{F})$ has two orbits $\mathfrak{X}^+$ and $\mathfrak{X}^-$. Let now $w=\begin{bmatrix} 0 & \mathfrak{p} \\ 1 & 0 \end{bmatrix}$; then w, considered as an element of $\mathbf{PGL}(2,\mathfrak{F})$, is of order 2: $w^2=\begin{bmatrix} \mathfrak{p} & 0 \\ 0 & \mathfrak{p} \end{bmatrix}\simeq I$. If α, $\beta\in\mathfrak{O}$,

$$\begin{bmatrix} 0 & \mathfrak{p} \\ 1 & 0 \end{bmatrix} \begin{bmatrix} \alpha \\ \beta \end{bmatrix} = \begin{bmatrix} \mathfrak{p}\beta \\ \alpha \end{bmatrix} \in \mathfrak{p}e_1\mathfrak{O} \oplus e_2\mathfrak{O},$$

and therefore $w(\mathcal{L}_0)\subseteq\mathcal{L}_1=\mathfrak{p}e_1\mathfrak{O}\oplus e_2\mathfrak{O}$.

Conversely $\begin{bmatrix} 0 & 1 \\ \mathfrak{p}^{-1} & 0 \end{bmatrix} \begin{bmatrix} \mathfrak{p}\alpha \\ \beta \end{bmatrix} = \begin{bmatrix} \beta \\ \alpha \end{bmatrix}$, and since $\begin{bmatrix} 0 & 1 \\ \mathfrak{p}^{-1} & 0 \end{bmatrix}\simeq w$ we have that $w(\mathcal{L}_1)\subseteq\mathcal{L}_0$. Thus, if Λ_0 is the class of $\mathcal{L}_0$ and Λ_1 the class of $\mathcal{L}_1$, we have that $w(\Lambda_0)=\Lambda_1$ and $w(\Lambda_1)=\Lambda_0$. Since $\mathbf{d}(\Lambda_0,\Lambda_1)=1$, we obtain that w is an inversion. We can now prove the following proposition.

(5.5) *PROPOSITION Let* t *and* s *be nonnegative integers such that* q+1=2t+s. *Then there exists a discrete subgroup* $\Gamma \subseteq \mathbf{PGL}(2,\mathfrak{F})$ *such that*

(i) Γ *acts faithfully and transitively on* $\mathfrak{X}$ *,*

(ii) Γ *is isomorphic to the free product of* s *copies of* $\mathbb{Z}_2$ *and* t *copies of* $\mathbb{Z}$ *,*

(iii) $\mathbf{PGL}(2,\mathfrak{F})=\Gamma.\mathbf{PGL}(2,\mathfrak{O})$.

PROOF. Observe that $\mathbf{PGL}(2,\mathfrak{F})$ acts transitively on $\mathfrak{X}$ and on Ω, and therefore it acts doubly transitively on $\mathfrak{X}$ and hence it acts transitively on the edges of $\mathfrak{X}$. But as we have just proved above $\mathbf{PGL}(2,\mathfrak{F})$ contains an inversion of order 2. It follows by (I,10.4) that $\mathbf{PGL}(2,\mathfrak{F})$ contains for every t and s with 2t+s=q+1 a faithful transitive subgroup satisfying (*i*) and (*ii*). As to (*iii*), we observe that

$$\mathbf{PGL}(2,\mathfrak{O}) = \mathbf{PGL}(2,\mathfrak{F})\cap K_{\Lambda_0} = \{g\in\mathbf{PGL}(2,\mathfrak{F}):\ g(\Lambda_0)=\Lambda_0\}.$$

If $g\in\mathbf{PGL}(2,\mathfrak{F})$ there exists one and only one $g'\in\Gamma$ such that $g(\Lambda_0)=g'(\Lambda_0)$. Thus $g^{-1}g'(\Lambda_0)=\Lambda_0$ and $g^{-1}g'\in\mathbf{PGL}(2,\mathfrak{O})$. Since $g=g'(g^{-1}g')^{-1}$, (*iii*) follows. ∎

We will now identify the subgroups $\mathbf{PGL}(2,\mathfrak{F})\cap G_\omega$ and $\mathbf{PGL}(2,\mathfrak{F})\cap G_\gamma$, where $\omega\in\Omega$ and $\gamma=(\omega,\omega')$ is a doubly infinite geodesic.

(5.6) *PROPOSITION.* (*i*) *Let* $\omega\in\Omega$; *then there exists a choice of basis* $\{e_1,e_2\}$ *in* V *with respect to which every element of* $\mathbf{PGL}(2,\mathfrak{F})$ *which fixes* ω *can be written as* $A=\begin{bmatrix} a & b\\ 0 & d\end{bmatrix}$, *with* $ad\neq 0$. *If, in addition,* A *is a rotation then we may also choose* $|a|=|d|=1$.

(ii) Let $\gamma=(\omega,\omega')$ *be a doubly infinite geodesic; then there*

exists a basis of V with respect to which every element A of $\mathbf{PGL}(2,\mathfrak{F})\cap G_\gamma$ *can be written as* $A=\begin{bmatrix} a & 0 \\ 0 & d \end{bmatrix}$ *(for $A\omega=\omega$ and $A\omega'=\omega'$) or* $A=\begin{bmatrix} 0 & b \\ c & 0 \end{bmatrix}$ *(when A interchanges ω and ω').*

PROOF. Let Λ_0 be a vertex of $\mathfrak{X}$ and let ω be identified by the infinite chain $\{\Lambda_0,\Lambda_1,\ldots,\Lambda_n,\ldots\}$. Let $\mathcal{L}_0\in\Lambda_0$ and $\mathcal{L}_n\in\Lambda_n$ be a maximal lattice satisfying $\mathcal{L}_n\subseteq\mathcal{L}_{n-1}$. Let $e_1\in\bigcap_{n=1}^{\infty}\mathcal{L}_n$, $e_1\neq 0$ and $\{e_1,e_2\}$ be a basis for V. If we write the elements of $\mathbf{PGL}(2,\mathfrak{F})$ as matrices with respect to the basis $\{e_1,e_2\}$, then the condition $A\in G_\omega$ implies that $A\begin{bmatrix}1\\0\end{bmatrix}=\begin{bmatrix}a\\0\end{bmatrix}$, in other words $A=\begin{bmatrix} a & b \\ 0 & d \end{bmatrix}$. Without loss of generality we may assume that Λ_n is the class of $\mathfrak{D}\oplus\mathfrak{p}^n\mathfrak{D}$ for every $n\in\mathbb{Z}$. Therefore it is easy to see that $\mathbf{PGL}(2,\mathfrak{F})\cap G_\omega\cap K_{\Lambda_0}$ is the set of matrices $\begin{bmatrix} a & b \\ 0 & d \end{bmatrix}$ with $|a|=|d|=1$ and $|b|\le 1$ while the set $\mathbf{PGL}(2,\mathfrak{F})\cap G_\omega\cap(K_{\Lambda_0}\setminus K_{\Lambda_{-1}})$ is the set of matrices $\begin{bmatrix} a & b \\ 0 & d \end{bmatrix}$ with $|a|=|b|=|d|=1$. Let $w=\begin{bmatrix} \mathfrak{p}^{-1} & 0 \\ 0 & 1 \end{bmatrix}$; $w(\Lambda_n)=\Lambda_{n+1}$ for every n, so w is a step-1 translation on the doubly infinite geodesic $\{\ldots,\Lambda_{-n},\ldots,\Lambda_{-1},\Lambda_0,\Lambda_1,\ldots,\Lambda_n,\ldots\}$. We have that $w^n\begin{bmatrix} a & b \\ 0 & d \end{bmatrix}w^{-n}=\begin{bmatrix} a & \mathfrak{p}^{-n}b \\ 0 & d \end{bmatrix}$. It follows that $\mathbf{PGL}(2,\mathfrak{F})\cap B_\omega\cap(K_{\Lambda_n}\setminus K_{\Lambda_{n-1}})$ is equal to the set of matrices $\begin{bmatrix} a & \mathfrak{p}^{-n}b \\ 0 & d \end{bmatrix}$ with $|a|=|b|=|d|=1$ while $\mathbf{PGL}(2,\mathfrak{F})\cap B_\omega\cap K_{\Lambda_n}$ is the set of matrices $\begin{bmatrix} a & b' \\ 0 & d \end{bmatrix}$ with $|a|=|d|=1$ and $|b'|\le q^n$. In particular $A\Lambda_n=\Lambda_n$ for every $n\in\mathbb{Z}$ if and only if $A=\begin{bmatrix} a & 0 \\ 0 & d \end{bmatrix}$ with $|a|=|d|=1$. This proves (*i*). To prove (*ii*), we observe that $\mathbf{PGL}(2,\mathfrak{F})$ acts doubly transitively on Ω, and therefore, as in part (*i*), without loss of generality we can suppose that $\gamma=\{\ldots,\Lambda_{-n},\ldots,\Lambda_{-1},\Lambda_0,\Lambda_1,\ldots,\Lambda_n,\ldots\}$. The fact that $A=\begin{bmatrix} a & b \\ c & d \end{bmatrix}$ fixes ω and ω' (that is the points $\begin{bmatrix}1\\0\end{bmatrix}$ and $\begin{bmatrix}0\\1\end{bmatrix}$ of the projective line) is equivalent to the fact that c=b=0. In part (*i*) we have proved that $A=\begin{bmatrix} a & 0 \\ 0 & d \end{bmatrix}$ fixes Λ_n for every n (that is A is a rotation of $\mathbf{PGL}(2,\mathfrak{F})\cap G_\gamma$ fixing ω and ω') if and only if $|a|=|d|=1$. This implies that if $A=\begin{bmatrix} a & 0 \\ 0 & d \end{bmatrix}$ and $|a|\neq|d|$ then A is a translation on the doubly infinite geodesic γ. The condition that A interchanges ω and ω' is equivalent to the fact that a=d=0. The matrices of this type are of order 2 and are exactly the rotations and the inversions of $\mathbf{PGL}(2,\mathfrak{F})\cap G_\gamma$ which interchange ω and ω'. It is easy to see that the matrix $A=\begin{bmatrix} 0 & b \\ c & 0 \end{bmatrix}$ is a rotation if and only if $\ln_q|b|-\ln_q|c|$ is even (and so A is an inversion interchanging ω and ω' iff $\ln_q|b|-\ln_q|c|$ is odd). This proves the proposition. ∎

REFERENCES

[A] F. Angelini, Rappresentazioni di un gruppo libero associate ad una passeggiata a caso, Tesi di Laurea in Matematica, Università di Roma "La Sapienza" 1989.

[B-K1] F. Bouaziz-Kellil, Représentations sphériques des groupes agissant transitivement sur un arbre semi-homogène, Bull. Soc. Math. France 116 (1988), 255-278.

[B-K2] F. Bouaziz-Kellil, Représentations sphériques des groupes agissant transitivement sur un arbre semi-homogène, Thèse de troisième cycle, Université de Nancy I, Nancy, 1984.

[BKu] H.Bass and R. Kulkarni, Uniform tree lattices, preprint.

[BP] W. Betori and M. Pagliacci, Harmonic analysis for groups acting on trees, Boll. U.M.I., (6) 3-B (1984), 333-349.

[BS1] C.Bishop and T. Steger, Three criteria for rigidity in PSL(2,ℝ), to appear in Bull. Amer. Math. Soc. .

[BS2] C. Bishop and T. Steger, Representation theoretic rigidity in PSL(2,ℝ), to appear in Acta Mathematica.

[BT] F. Bruhat and J. Tits, Groupes réductifs sur un corps local, Publ. Math. I.H.E.S. n. 41.

[C1] P. Cartier, Géométrie et analyse sur les arbres, Sem. Bourbaki 24 1971/72, exposé 407, Lecture Notes in Math. 317, 123-140, Springer, 1973.

[C2] P. Cartier, Fonctions harmoniques sur un arbre, Symp. Math. 9 (1972), 203-270.

[C3] P. Cartier, Harmonic analysis on trees, Proc. Symp. Pure Math. A.M.S. 26 (1972), 419-424.

[C F-T] C. Cecchini and A. Figà-Talamanca, Projections of uniqueness for $L^p(G)$, Pacific J. of Math. 51 (1974), 37-47.

[Ch] F. Choucroun, Groupes opérant simplement transitivement sur un arbre homogène et plongements dans $PGL_2(k)$, C. R. Acad. Sc. Paris Ser. A 298 (1984), 313-315.

[CdM] J. M. Cohen and L. de Michele, Radial Fourier-Stieltjes algebra on free groups, Contemporary Math. 10 (Operator algebra and K-Theory), 1982.

[CM] M. Culler and J. Morgan, Group actions on $\mathbb{R}$-trees, Proc. London Math. Soc. 55 (1987), 571-604.

[CS] M. Cowling and T. Steger, Irreducibility of restrictions of unitary representations to lattices, to appear in J. Reine Angew. Math. .

[D1] J. Dixmier, Les C^*-algèbres et leurs représentations, Gauthier-Villars, Paris, 1969.

[D2] J. Dixmier, Les algèbres d'opérateurs dans l'espace hilbertien, Gauthier-Villars, Paris, 1969.

[Di] G. van Dijk, Spherical functions on the p-adic group PGL(2), Indag. Math. 31 (1969), 213-241.

[DS] N. Dunford and J. T. Schwartz, Linear operators, Part II, Interscience, 1963.

[F] J. Faraut, Analyse harmonique sur les paires de Gelfand et les espaces hyperboliques, Analyse Harmonique Nancy 1980, CIMPA, Nice 1983.

[FP] J. Faraut and M. A. Picardello, The Plancherel measure for symmetric graphs, Ann. di Mat. Pura Appl. (IV) 138 (1984), 151-155.

[F-T P1] A. Figà-Talamanca and M.A. Picardello, Spherical functions and harmonic analysis on free groups, J. Funct. Anal., 47 (1982), 281-304.

[F-T P2] A. Figà-Talamanca and M.A. Picardello, Harmonic analysis on free groups, Lectures Notes in Pure and Applied Mathematics, Marcel Dekker, New York, 1983.

[F-T P3] A. Figà-Talamanca and M. A. Picardello, Restriction of spherical representations of $PGL_2(\mathbb{Q}_p)$ to a discrete subgroup, Proc. Amer. Math. Soc. 91 (1984), 405-408.

[F-T S1] A. Figà-Talamanca and T. Steger, Harmonic analysis on trees, Symp. Math. 29 (1987), 163-182.

[F-T S2] A. Figà-Talamanca and T. Steger, Harmonic analysis for anisotropic random walks on homogeneous trees, Memoirs Amer. Math. Soc., to appear.

[G] S. A. Gaal, Linear analysis and representation theory, Springer-Verlag, Berlin Heidelberg New York, 1973.

[GGP] I.M. Gel'fand, M.I. Graev and I.Pyatetski-Shapiro, Theory of representations and automorphic functions (Generalized functions, vol. 6, in Russian) (Moscow,1966).

[Gu] C. L. Gulizia, Harmonic analysis of SL(2) over a locally compact field, J. Funct. Anal., 12 (1973), 384-400.

[H] U. Haagerup, An example of a nonnuclear C^*-algebra which has the metric approximation property, Invent. Math. 50 (1979), 279-293.

[He] S. Helgason, Eigenspaces of the Laplacian, integral representations and irreducibility, J. Funct. Anal. 17 (1974), 328-353.

[Hz1] C. Herz, Le rapport entre les algèbres A_p d'un groupe et d'un sous-groupe, C.R. Acad. Sci. Paris A 271 (1970), 244-246.

[Hz2] C. Herz, Synthèse spectrale pour les sous-groupes par rapport aux algèbres A_p, C. R. Acad. Sci. Paris A 271 (1970), 316-318.

[I] I.M. Isaacs, Character theory of finite groups, Pure and Applied Mathematics, Academic Press, New York, 1976.

[JL] H. Jacquet and R. P. Langlands, Automorphic Forms on GL(2), Lect. Notes in Math. 114 Springer-Verlag, 1970.

[K1] S. Kato, Irreducibility of principal series representations for Hecke algebras of affine type, J. Fac. Sci. Univ. Tokyo Sect. I A Math. 28 (1981), 929-943.

[K2] S. Kato, On eigenspaces of the Hecke algebra with respect to a good maximal compact subgroup of a p-adic reductive group, Math. Ann. 257 (1981), 1-7.

[Ko] N. Koblitz, p-adic Numbers, p-adic Analysis and Zeta-Functions, Graduate Texts in Math. n. 58, Springer-Verlag 1977.

[KP] A. Korányi and M. A. Picardello, Boundary Behaviour of Eigenfunctions of the Laplace Operator on Trees, Ann. Scuola Norm. Sup. Pisa Ser. IV vol. 13 n. 3 (1986), 389-399.

[KPT] A. Korányi, M. A. Picardello and M. H. Taibleson, Hardy spaces on non-homogeneous trees, harmonic analysis, symmetric spaces and probability theory, Cortona (Italy) 1984, Symp. Math. 29 (1987), 205-254.

[KS1] G. Kuhn and T. Steger, Restrictions of the special representation of $Aut(Tree_3)$ to two compact subgroups, Rocky Mountain J., to appear.

[KS2] G. Kuhn and T. Steger, A characterization of spherical series representations of the free group, Proc. of the Amer. Math. Soc., to appear.

[L] S. Lang, $SL_2(\mathbb{R})$, Addison-Wesley, Reading, Mass. , 1975.

[Lu] A. Lubotzky, Trees and discrete subgroups of Lie groups over local fields, Bull. of the Amer. Math. Soc. 20 (1989), 27-30.

[M] G. W. Mackey, Induced representations of locally compact groups I, Ann. Math. vol. 55 (1) 1952, 101-139.

[MM] J. McMullen, Lecture Notes of the Dep. of Pure Math. of the University of Sydney on a paper of Ol'shianskii.

[MZ1] A. M. Mantero and A. Zappa, The Poisson transform on free groups and uniformly bounded representations, J. Funct. Anal. 51 (1983), 372-399.

[MZ2] A. M. Mantero and A. Zappa, Special series of unitary representations of groups acting on homogeneous trees, J. Austral. Math. Soc. (Series A) 39 (1985), 1-6.

[N1] C. Nebbia, Groups of isometries of trees with simply transitive subgroups, Boll. U.M.I. (7) 1-A (1987), 387-390.

[N2] C. Nebbia, Groups of isometries of a tree and the Kunze-Stein phenomenon, Pacific J. of Math. vol. 133 (1) (1988), 141-149.

[N3] C. Nebbia, Amenability and Kunze-Stein property for groups acting on a tree, Pacific J. of Math., vol. 135 (2) (1988), 371-380.

[N4] C. Nebbia, Classification of all irreducible unitary representations of the stabilizer of the horicycles of a tree, Israel J. of Math., vol.70 n.3 (1990), 343-351.

[N5] C. Nebbia, A note on the amenable subgroups of PSL(2,$\mathbb{R}$), Mh.Math. 107 (1989), 241-244.

[O1] G. I. Ol'shianskii, Representations of groups of automorphisms of trees, Usp. Mat. Nauk, 303 (1975), 169-170.

[O2] G. I. Ol'shianskii, Classification of irreducible representations of groups of automorphisms of Bruhat-Tits trees, Functional Anal. Appl. 11 (1977), 26-34.

[O3] G. I. Ol'shianskii, New "large" groups of Type I, Journ. Sov. Math. 18 (1982), 22-39.

[P] T. Pytlik, Radial functions on free groups and a decomposition of the regular representation into irreducible components, J. Reine Angew. Math. 326 (1981), 124-135.

[PW] M. A. Picardello and W. Woess, Martin boundaries of random walks: ends of trees and groups, Trans. of the A.M.S. vol. 302 (1) (1987), 185-205.

[R] W. Rudin, Real and complex analysis, McGraw-Hill, New York, 1966.

[S] E. Seneta, Non-negative Matrices and Markov Chains, Springer-Verlag, Berlin, 1973.

[Se] J. P. Serre, Arbres, Amalgames, SL_2, Astérisque 46, 1977.

[Si] A. J. Silberger, PGL_2 over the p-adics: its representations, spherical functions and Fourier analysis, Lect. Notes in Math. 166 Springer-Verlag, 1970.

[SS] P. J. Sally Jr. and J. A. Shalika, The Plancherel formula for SL(2) over a local field, Proc. N.A.S. vol. 63 (1969), 661-667.

[St1] T. Steger, Finite reducibility of random walk representations of free groups, in preparation.

[St2] T. Steger, Restriction to cocompact subgroups of representations of Aut(Tree), Lectures given at the University of Rome "La Sapienza", June 1987.

[St3] T. Steger, Restriction to cofinite Fuchsian groups of representations of SL(2,$\mathbb{R}$), Lectures given at the University of Rome "La Sapienza", June 1986.

[T] A. E. Taylor, Introduction to Functional Analysis, J. Wiley and Sons Inc., New York, 1958.

[Ti1] J. Tits, Sur le groupe des automorphismes d'un arbre, Essays on topology and related topics, Mémoires dédiés à G. de Rham, Springer-Verlag, Berlin 1970, 188-211.

[Ti2] J. Tits, A "theorem of Lie-Kolchin" for trees, Contribution to algebra, A collection of papers dedicated to Elli Kolchin, Academic Press 1977, 377-388.

[Y] K. Yosida, Functional analysis, Fourth Edition, Springer-Verlag, 1974.

[W] A. Weil, Basic number theory, Springer-Verlag, Berlin Heidelberg New York, 1973.

SYMBOLS

INDEX